EXTREME SCIENCE

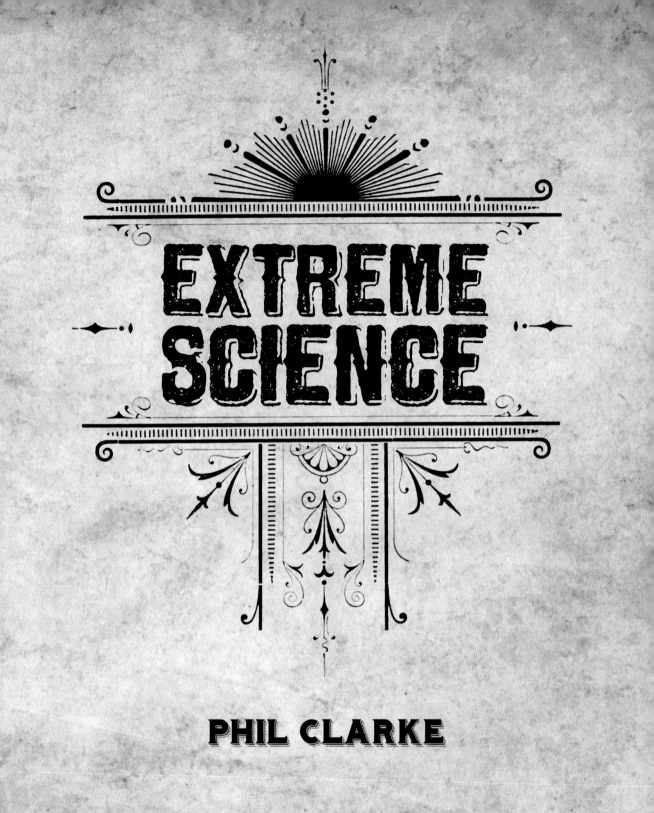

EXTREME SCIENCE

PHIL CLARKE

CHARTWELL
BOOKS, INC.

CONTENTS

INTRODUCTION

Have you ever wanted to look beyond the clouds and the stars, or to know what causes the trees to bud? And what changes the darkness into light? But if you talk like that, people call you crazy. Well, if I could discover just one of these things, what eternity is, for example, I wouldn't care if they did think I was crazy.

Frankenstein (1931 movie)

We are all familiar with the classic stereotype of the mad professor; a white skinned, white haired, wide eyed, bespectacled old man in a white coat, whose laboratory is crammed full of smoking test tubes, putrid Petri dishes, bubbling potions, and mysterious, groaning creations shrouded in dust sheets. It is reinforced on a daily basis by makers of cartoons, comic strips, movies, TV and websites the world over, but where exactly did it originate from? Who was the first mad scientist?

A handful of individuals, both historical and fictional, spring to mind. Among these, perhaps the most notorious is Dr Victor Frankenstein, but the stereotype itself is actually infinitely older than Mary Shelley's novel of 1818. Scientists have been portrayed in this way for as long as science has existed, beginning in Ancient Greece with the mythical architect and inventor, Daedalus. We don't know if he had white hair and he almost certainly didn't wear a white coat, but his story bears all the hallmarks of the genre. A cloistered genius who, driven mad from loneliness and frustration, creates a machine that causes more problems than it solves.

Daedalus and Icarus

According to legend, Daedalus was employed by King Minos of Crete to construct a labyrinth in which to house the half-man and half-bull Minotaur. When the labyrinth was finished, King Minos was worried that Daedalus might reveal its secrets to the world, and so he imprisoned Daedalus in a tall tower along with his young son, Icarus.

Desperate for freedom, Daedalus studied birds in flight and decided that he could make Icarus a pair of wings out of wax and feathers, so that he might escape. He warned his son to be careful with his gift. But, once Icarus was in the air, he enjoyed it so much that he grew too confident, and flew too close to the sun. The sun melted the wax that held the delicate wings together and they began to fall apart. Poor Icarus plummeted into the sea, where he drowned.

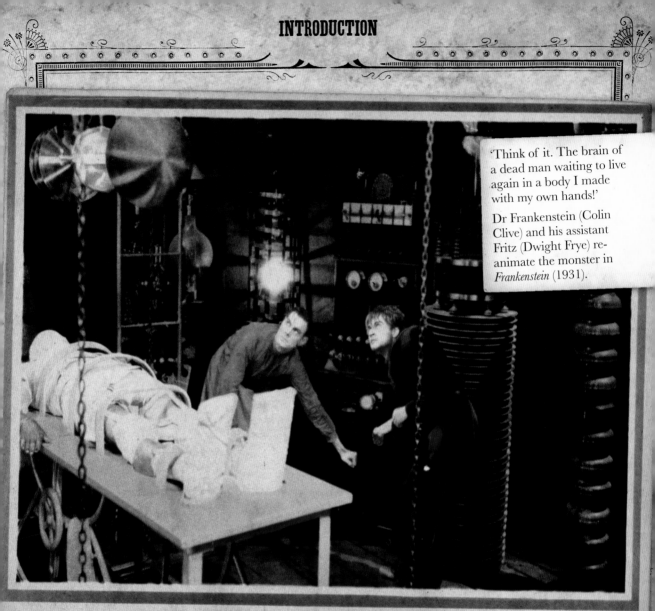

'Think of it. The brain of a dead man waiting to live again in a body I made with my own hands!'

Dr Frankenstein (Colin Clive) and his assistant Fritz (Dwight Frye) re-animate the monster in *Frankenstein* (1931).

CARL LAEMMLE
presents

"FRANKENSTEIN"
THE MAN WHO MADE A MONSTER

A UNIVERSAL
PICTURE

In some ways it is possible to see the Icarus story as a precursor to Mary Shelley's *Frankenstein*. After all, they both warn against messing with things you don't understand. Most people avoid the unfamiliar in everyday life, but for a scientist, it is as if they have no choice but to dabble, because they regard the alternative of ignorance as a fate worse than death. It is this singularity of vision that sets them apart from the rest of society but it also encourages us to categorize them as insane.

The Alchemists

The public perception of the 'mad scientist' is very closely related to that of the alchemist. Alchemists were prototype chemists, but as well as chemistry they concerned themselves with mythology, spirituality and religion. Many were famously obsessed with the idea of turning base metals into gold. These men often did act strangely because they were exposed to dangerous chemicals, the properties of which were not fully understood during their lifetimes.

The famous scientist, Sir Isaac Newton was fascinated by alchemy, and is said to have suffered from mercury poisoning as a result of his experiments in the field, or his work as warden of the Mint. His symptoms included acute irritability, chronic insomnia and hyperactivity – all qualities we've come to associate with the typical 'boffin'.

But the alchemists of the Middle Ages were by no means alone; history offers us plenty more examples of scientists whose work bought them into contact with substances which, ultimately, destroyed them. Just look to William Halsted; the promising and brilliant American surgeon who set out to prove that cocaine could be used as an effective local anaesthetic, and wound up battling a savage addiction for the rest of his life. Halsted, and others like him, simply accepted madness and premature death as an occupational hazard.

Franz Reichelt

Perhaps the most brave and dedicated men (and women), of science are the ones who are prepared to make the ultimate sacrifice in their pursuit of knowledge. Although whether they are remembered as heroically brave or dangerously reckless, depends wholly on the outcomes of their research. Franz Reichelt was a tailor from Vienna who developed a suit for aviators that converted into a parachute, so that they could survive a fall if forced to eject from an aircraft.

Having tested a few prototypes (using dummies) from the fifth floor of his apartment building, he felt he needed to experiment using a higher launching platform, so he petitioned the Parisian Prefecture of Police for permission to conduct a test from the Eiffel Tower. Eventually they agreed, but they failed to realize that, this time, rather than use a dummy, Reichelt intended to jump himself.

On Sunday 4 February 1912, with a crowd of friends, journalists, press photographers and cinematographers waiting eagerly beneath him, Reichelt launched himself from the first platform of the Eiffel Tower. Unfortunately, his parachute suit failed to deploy and he hurtled straight to the ground. He was killed instantly.

Perhaps Reichelt was foolish to trust his life entirely to his invention, but surely, had it worked, he would be remembered, not as an idiot who leapt to his death wearing a glorified shell-suit – but as a man who risked his own life in a bid to save countless others. As it happened, he succeeded only in radically shortening his own life and, arguably, inventing the extreme sport of base jumping.

Only 12 months later, Stefan Banic successfully demonstrated his parachute design by jumping from a 15 storey building in Washington DC. He donated his design to the US military and is now credited with saving scores of US aviators during World War I.

A Question of Ethics

Of course, not every scientist is equipped to act as their own guinea pig, and when that's the case either animals or humans (often both) are drafted in to fill this gap. In many cases this practice has widened the divide between the scientific community and the rest of society, and added to the image of the scientist as someone who operates beyond conventional notions of morality.

In 2007, the eminent British professor, Sir David King, outlined a seven-point Universal Ethical Code for Scientists. A modern-day scientist's equivalent of the Hippocratic Oath, he believes that, if all scientists were to follow it, society would place more trust in them. Notice that the code only asks scientists to *minimize* negative impact on people, animals and the environment.

THE CODE

- Act with skill and care, keep skills up-to-date
- Prevent corrupt practice and declare conflicts of interest
- Respect and acknowledge the work of other scientists
- Ensure that research is justified and lawful
- Minimize impacts on people, animals and the environment
- Discuss issues science raises for society
- Do not mislead, present evidence honestly

Wouldn't it be interesting if we could find out what men like the proto-hypnotist Franz Mesmer, the vomit drinking Stubbins Ffirth or the electrocutioner Luigi Galvani thought of his proposal? Would they have been in favour of such a thing? Would their work have fallen within its rules?

Certainly, there are a handful of people in this book whose research would surely have come apart at the seams if they'd been expected to adhere to a formal code of ethics. There are also a few who believed that they were above any rule of law, and would have been quite happy to continue regardless.

The fourth and the fifth points of the code, in particular, could potentially have troubled the men and women of MKULTRA, who continued to spike unsuspecting civilians with LSD for years. Not content with torture, murder and battery, they even opened and ran a chain of state-sponsored brothels as a front for their covert operations. In this day and age, it's difficult to believe that such actions could ever be justified.

This book delves into the lives of people and organizations that have crossed boundaries in search of knowledge, often going beyond what's considered right, sensible or even sane, in pursuit of the truth. With that in mind, be warned that much of the content is shocking in nature.

Many of the men and women written about within these pages have spent their careers performing experiments on animals and humans alike, with varying degrees of success. In some cases, they have gone beyond what you could reasonably call the interests of society, and inflicted suffering just for the sheer thrill of it. And so it goes without saying that some sensitive readers may find these stories, and the accompanying images, disturbing. If however, you are of an appropriate age and attitude, read on.

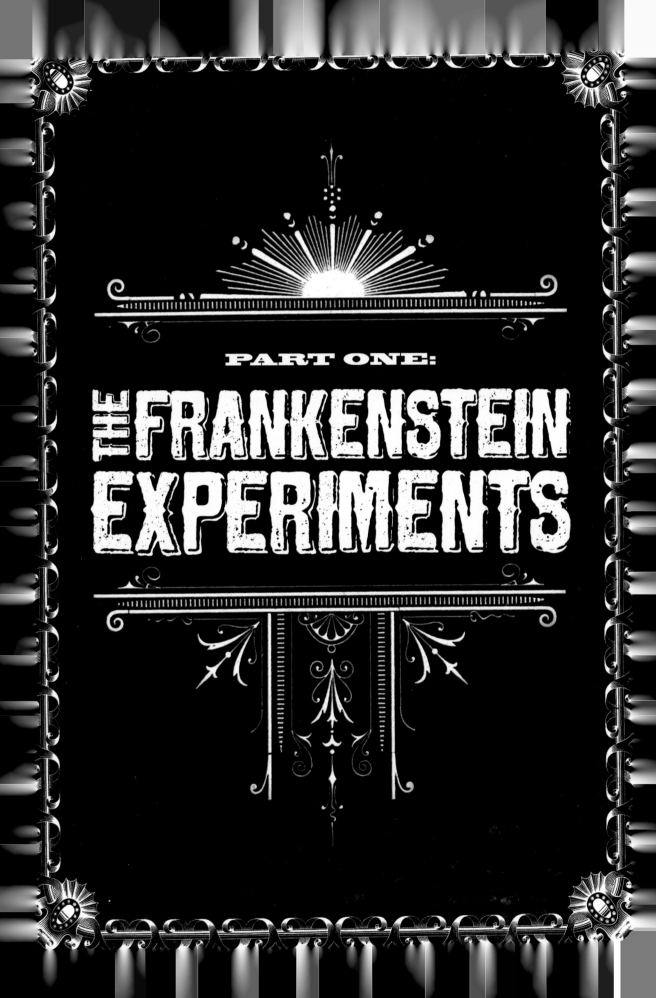

PART ONE:

THE FRANKENSTEIN EXPERIMENTS

THE ELIXIR

Johann Konrad Dippel (1673 – 1734). Believed by many to be the original Doctor Frankenstein, Johann Konrad Dippel led a wandering life, escaping persecution in every province thanks to his theological views. With ambitions for acquiring everlasting life, this nomadic nostrum maker dabbled in the dark art of alchemy, distilling the blood and bones of animal cadavers in search of an elixir vitae. Yet it was to be his serendipitous success in colour creation that would guarantee his immortality.

Born Under a Bad Sign in Castle Frankenstein

Deep in the heart of the Hessian Rhineland Lutheran pastor Johann Philipp Dippel and his wife Anna became victims of religious persecution and, in 1670, were forced to seek refuge at a hilltop fortress overlooking the vast Odenwald. This dark and brooding place went by the name of Castle Frankenstein and three years later, behind its Stygian walls, they bore a son whom they named Johann Konrad.

From a very early age the boy was schooled by his father to follow the paternal career path and become the fifth clergyman in the family. An exceptionally bright child, Johann Junior was quick to comprehend religious matters and at the tender age of nine revealed a contrary side to his character; openly criticizing the question and answer format of the Catechism. He also convinced himself his superior intellect was a divine gift, believing he was ordained as a prophet by God. Such delusions of grandeur would become an indelible feature of Dippel's career and life as a whole.

In May 1691 Johann enrolled at the University of Giessen and during his two-year stay rapidly achieved a reputation as an ardent defender of orthodox Lutheranism. Gaining his master's degree in theology he began looking for a professorship at his own university and those nearby. Unable to secure an academic post, Dippel moved to Strasbourg in the summer of 1695 and immersed himself in the writings of his religious opponents, the Pietists. It was during this time that he began a remarkable volte-face in his beliefs.

Troublesome Tracts

While lecturing on astrology and palmistry at the university of Strasbourg, Johann's theological stance started to lean in favour of the Pietists who, in direct conflict with the Lutheran Orthodoxists, rejected the idea that church ritual and institutionalized dogma was the way to heaven. Their tenets advocated deeper personal relationships with the spiritual world.

Adopting this new doctrine, Dippel began preaching its principles about town so vociferously that he became embroiled in a bitter dispute. Unable to come to terms a duel was declared to settle the matter and, following the requisite swords at dawn, Johann killed his opponent forcing him to flee the free city of Strasbourg in the summer of 1696.

Several months later he had settled back in Giessen and, whilst serving as the personal tutor to the six-year-old son of the Landgrave of Hesse, he persisted with his polemical outbursts. Encouraged by fellow Pietists such as Gottfried Arnold, Dippel set about putting his beliefs down on paper. Over the next two years he wrote over 14 inflammatory tracts under the pseudonym Christianus Democritus, his views subjecting him to persecution from the orthodox church.

Dippel's Oil

Excluded from both the academic and ecclesiastical worlds, Johann Dippel now turned his considerable mind to a serious study of alchemy. This hermetic science fixated with the transmutation of base metals into gold and the existence of an *elixir vitae* was all the rage at the turn of the century and its allure quickly consumed the theologian. Poring over the alchemical texts of Ramon Llull, a Majorcan mystic, he confidently predicted that gold-making was not beyond his abilities. So confident was he of success that he

purchased a small estate on credit.

Eight months of perpetual heating of metals brought nothing more than a cracked crucible so the deluded Dippel turned his hand to creating the other element of alchemy: the elixir of life. He studied the common practice of extracting ammonia from hartshorn and applied this approach to the blood, bones and unwanted remains of animal carcasses. By liquidizing this mess and distilling it through iron tubing he arrived at his end product: Dippel's Oil.

This foul-smelling dark-coloured distillate of carbonized bone and blood was not however the alchemists' dream of an immortalizing nostrum; Dippel had failed once again. Not wanting to admit defeat, he claimed this noisome liquid was a universal remedy capable of curing all colds, fevers and even epilepsy. Believing that anything so distasteful had to be beneficial, customers came from far and wide to partake of the panacea. Swiftly however its inefficacy became apparent, though it did acquire some legitimate uses; as an ingredient in sheep dip, and as an animal repellent, and was sold well into the 19th century.

A Discovery Worth Dyeing For

Sales in the all-curing oil were not enough to balance the books so in 1704 the debt-ridden Dippel left his home-town and moved to Berlin. It was here in a shared laboratory that the foul tonic led to a further discovery, thanks largely to his lab partner. One day while preparing a batch of crimson, artists' colour manufacturer Heinrich Diesbach discovered he had run out of an essential ingredient: potassium carbonate. Borrowing some from Dippel, the pigment maker added the potash to the mixture of crushed cochineal insects and assorted chemicals.

Unbeknownst to the German colourist, this stock had been used in the alchemist's creation of his animal oil and had become contaminated with cyanides. Consequently, as Diesbach stirred his concoction he witnessed a surprise development. Rather than turning a deep red as expected, the preparation produced a vivid blue.

When Dippel saw what had occurred he set to work on recreating the colour. Further tests with the cyanide-laden potash led to identical batches of blue pigment.

This was a real find. To date, the colour had been expensive to manufacture, requiring the semi-precious stone *lapis lazuli* retrieved from mines of Badakhshan. Now it was possible to produce it cheaply. Initially called Berlin Blue, one of its first uses was to dye the uniforms of the Prussian army, from which it acquired its more common name: Prussian Blue.

Dippel the Drifter

Three years after his brush with colour creation, Dippel was driven out of Berlin. Various high-profile officials had grown tired of his continual anti-orthodox views

Castle Frankenstein, The Odenwald Valley, Germany.
Mary Shelley always said the name 'Frankenstein' came to her in a dream. But legend has it, that while travelling through Germany, Mary and her husband, Percy Bysshe Shelley, visited Castle Frankenstein in the Odenwald Valley, where Konrad Dippel had experimented on human corpses. This visit almost certainly inspired her famous novel of 1818.

on religion forcing him to escape to Frankfurt via Koestritz. Lying low and away from the church's wrath for several years, Johann eventually moved to Leyden where he obtained his doctorate in medicine in 1711. Buying a house outside Amsterdam and practising as a bona fide physician he seemed to be settling down and a period of relative calm ensued.

It did not take long for Dippel to court further controversy, however. Unable to keep his Pietist faith under wraps, he published yet another antagonistic pamphlet which promptly led to his exile from Holland. Next stop was Sleswig-Holstein but yet again his outspoken criticism of the church aroused the hatred of the Danish courtiers and in 1719 he was condemned to permanent imprisonment on the island of Bornholm in the Baltic Sea.

Luckily for Dippel, he served just seven years of his eternal sentence and was released in 1726 at the request of the Queen of Denmark. Ever the wanderer, the ex-convict moved on a year later. His pungent oil had attracted the attention of Swedish royalty and he was asked to become the physician to King Frederick I. Besieged by alchemists, the monarch gave Dippel the responsibility of unmasking the pretenders who promised him limitless wealth. Such close contact with these occultists only enhanced his desire to discover alchemical treasure.

Life Neverlasting

Unsurprisingly, Dippel soon ruined his standing with the Swedish court, publishing his final heretical work in 1729. Banished from another country, he returned to his native land once more to devote his time to refining his *elixir vitae*. Reports state he took up residence in Liebenburg, a small village in Lower Saxony, where he worked feverishly in secret to create a genuine rejuvenating potion.

Some time later, Dippel emerged from his laboratory believing he had succeeded where others had failed. Calling it his *Arcanum Chymicum* or secret chemical, the half-crazed Hessian approached the local landgrave offering the dubious elixir in exchange for the feudal rights to his birthplace, Castle Frankenstein. His proposal was rejected. However, myths persist to this day that Johann did stay in the castle, performing gruesome experiments on cadavers exhumed from nearby graveyards.

The legend continues that the local villagers soon caught wind of his despicable deeds – including the re-animation of corpses – and, in true Frankenstein style, ran the monster out of town. While no real evidence exists to place him here, Dippel is believed to have resided in another castle some 100 miles north. At the behest of a benevolent count, the deranged doctor was invited to stay at Castle Wittgenstein and was provided with laboratory space in nearby Berleburg.

Hiding away practising his dark arts only helped to perpetuate the rumours that he had made a Faustian pact with the Devil, relinquishing his soul in exchange for alchemical knowledge. Despite claims that he had successfully made an elixir which would ensure he lived until the year 1808, Johann Konrad Dippel passed away inside the castle on 25 April 1734. Rumour has it he was poisoned by the very potion that he believed guaranteed his immortality.

GRANDFATHER OF EVOLUTION

Erasmus Darwin (1731–1802). Now honoured for his forward-thinking theories on evolution, this polymath poet suffered for his science, ridiculed by the powers-that-were who saw him as a subversive shadow rather than a leading light in the Enlightenment era. The truth was that Erasmus Darwin was decades and even centuries ahead of his time. Many of his futuristic ideas gradually came to fruition and his evolutionary beliefs provided the foundation from which his grandson, Charles Darwin, would eventually discover how mankind came to be.

Early Erasmus

Born on 12 December 1731 at Elston Hall, a substantial mansion in the Nottinghamshire countryside, Erasmus Darwin became the seventh child of a retired barrister and his pigeon-fancying wife. As an inquisitive boy growing up on the vast estate he quickly developed a passion for science, choosing to perform crude experiments with clocks and other mechanical devices.

He was happiest inside tinkering with machinery, even dabbling with electricity. And perhaps safer. He nearly drowned while on a fishing trip with his older brothers; his spiteful siblings choosing to tie him inside a sack and throw him into the river.

Physical exercise, then, never captured his curiosity. But when it came to employing his mind, Erasmus showed himself to be a keen and avid student. Spending nine years at Chesterfield Grammar School he excelled, devouring the classical and literary education it offered. He then enrolled at St John's College, Cambridge in June 1750 where he was allowed to develop his scientific mind. Visits to anatomy lectures in London and further learning at Edinburgh Medical School soon led in the summer of 1756 to his achieving a doctorate.

Erasmus Darwin, writer, scientist and grandfather of Charles Darwin.

Doctor Darwin

Immediately following his graduation Erasmus moved to Nottingham, ready to begin his fledgling career as a physician. However, his practice failed to thrive. This was due in no small degree to one particular patient who, following a fight, suffered a fatal infection from his wound. Darwin's inability to prevent death caused the locals to lack confidence in him and the community chose to seek medical assistance elsewhere.

His medical judgement discredited, Doctor Darwin moved to Lichfield in Staffordshire where his fortunes changed almost overnight. Mr Inge, a young gentleman of means, lay at death's door racked with a terrible fever, his personal physician resigned to losing his patient. Desperate to save her son, Inge's mother contacted the newly-arrived doctor who managed to restore the young man's health. No record exists of the treatment Erasmus gave his first patient, but the successful cure gave him some welcome publicity throughout the cathedral city.

Backed by influential patrons, his practice blossomed. Soon he was attending to an exclusive clientele of local gentry, titled gentlemen and land-owning families. Travelling thousands of miles a year in his personal carriage, he quickly built up an unequalled reputation as one of the most attentive doctors in the West Midlands. His talents at this stage lay not with a magic potion or unique salve – in fact his treatments followed the opium-prescribing habits of a typical Georgian physician. Where he succeeded was in his bedside manner, winning the confidence of the sick through sympathy and care.

The Lunarticks

As his professional reputation grew so did his popularity in social circles. Blessed with a gregarious nature and, despite suffering an acute stammer, a cutting wit, Erasmus built up a vast network of associates. This led him to establish The Lunar Society in the 1760s, a loosely-knit social club for intellectuals and industrialists. To partake in sophisticated discussion they met monthly on a Monday afternoon nearest the full moon, hence the name. This was not for any esoteric belief in werewolves but for the simple purpose that it afforded them the most light to see their way home after their enlightened exchanges.

The Lunarticks, as they were called (a pun on the two words 'lunar' and 'lunatic'), conversed on a variety of subjects, suiting Darwin's polymath tendencies. Geology to geometrics, meteorology to mechanics, this smart set – while less prestigious than the Royal Society – became the doctor's preferred crowd. And as the membership increased The Lunarticks quickly became a driving force behind the Industrial Revolution in England.

These moonlit meetings helped Erasmus to form some new ideas, not least on evolution. Following the unearthing of some fossils in the area, he surmised that perhaps animal life may have developed through the ages. To this end, he added a motto to the three scallop shells on his family coat-of-arms: *Omnia ex conchis* or Everything From Shells. This he had printed on his personal coach and book-plates, which incurred the wrath of the local clergy who accused him of renouncing God, forcing him to repaint his personal coach.

Commonplace Creations

Possessing a vast scope of interests, Darwin was able to turn his mind to other matters. The pages of his Commonplace Book, a journal favoured by those of the Enlightenment era to document ideas, was packed with innovative designs and sketches. Yet there was nothing commonplace about them. Inside were inventions centuries ahead of their time. Erasmus created workable blueprints for a vast array of devices, ranging from the photocopier and the loudspeaker to the automatic flushing toilet.

His fertile mind left no field of endeavour untouched by his creative genius. Darwin sent countless papers to the *Philosophical Transactions*, a scientific journal published by the Royal Society, and developed a number of mechanical apparatus for his Lunartick friends. Believing his ideas would be in direct conflict with his role as a physician, he refused to patent any preferring others to benefit. Like the horizontal windmill designed for Josiah Wedgwood, allowing the pulverizing of ceramic materials whichever way the wind blew.

Erasmus also imagined one of the first feasible rocket engines. His 'fiery chariot' consisted of two separate tanks of hydrogen and oxygen plumbed to a combustion chamber. An advanced version of this primitive propellant-induced machine would eventually be used in the NASA space launches of the 1960s. Visionary concepts such as this were born from his need to create. As Darwin himself put it: 'A fool is a man who never tried an experiment in his life.'

The Force Is With You

Darwin's involvement in mechanical science did not however hinder his primary work as a physician. And even in this capacity, Erasmus stood at the cutting edge. The in-vogue marvel of electricity was fast becoming a potential panacea for all ills and Georgian scientists toyed with electrostatic devices and Leyden Jars hoping to harness the powerful qualities of the shocking wonder. And the inventive Doctor Darwin was no different. He believed all life was driven by an ethereal energy which he called spirit of animation. This spirit shared similarities with electricity and their connection caused Erasmus to suggest a galvanic charge could be beneficial in treating the body and mind.

His medical practice gave him ample opportunity to test out his theory on a variety of ailments. Doctor Darwin visited a maltman suffering from a tapeworm. After discovering the usual treatment - a mix of tin and quicksilver washed down with a cathartic of sodium sulphate – was powerless to shift the parasite, he administered 20 short sharp shocks from a Leyden bottle through the stomach. Several sessions later and the worm was successfully purged.

With such a potent tool in his medical bag, Erasmus sought to apply the electrical elixir to more disorders, including hydrocephalus, jaundice, toothache and cramps. It was these electro-therapeutic treatments along with Darwin's other experiments in galvanism that provided Mary Shelley with vital inspiration for writing her *Frankenstein*. Word of his medical expertise reached not only the ears of authors but also the King himself. After curing a cousin of the Royal

Household, George III offered him the role of royal physician, which he declined on numerous occasions.

Poetic Justice

Erasmus was not just an inspiring force for writers, he was a writer himself. From a young age he dabbled in poetry but kept these artistic talents secret during his early years as a physician, unwilling to risk jeopardizing his career. After translating Linnaeus' plant classification from Latin throughout the 1780s a burgeoning passion for botany led him to write *The Botanic Garden*; two lengthy didactic poems popularizing the earlier translation. So critically-acclaimed were they that following their publication Darwin was hailed as the leading English poet of the time, respected by the likes of Coleridge and Wordsworth.

His next work was, scientifically, his most important. A practical textbook on medicine and physiology, *Zoonomia* was written in two volumes and included a comprehensive catalogue of nearly 500 diseases along with suitable cures. However, tucked away in a latter chapter, it contained his theory of biological evolution, stating: *all warm-blooded animals have arisen from one living filament.*

This radical hypothesis saw him displease both the state and the church. Britain was now at war with France and a mood of anti-intellectualism had descended upon the country in a bid to maintain order. Free thinkers were ejected from the land, transported to Australia; Joseph Priestley, a fellow Lunar man, had his Birmingham home destroyed by the mob. Now they came after Erasmus. Seen as a subversive element, his work was lampooned by satirists. Cast as the twisted mad scientist maniacally searching for his ethereal fire inside his laboratory, Darwin gradually lost his reputation.

Beaten back by the establishment, Erasmus retired his practice in 1801. His unconventional beliefs persisted until the end. Thinking heavy eating promoted good health he consumed extreme amounts and grew to such a size that he required a semi-circle cut out of his dining table to accommodate his bulging waistline. Weeks after moving into Breadsall Priory, Erasmus Darwin died suddenly from a lung infection on 18 April 1802.

A year later his final poem, *The Temple of Nature*, was posthumously published focusing on his evolutionary ideas. These thoughts were eventually passed down to his grandson, Charles Darwin, who would famously prove the veracity of his grandfather's theory. It was a long time coming but in the end this largely forgotten physician had the last laugh; his beliefs in biological development affirmed and his reputation restored.

Organic life beneath the shoreless waves
Was born and nurs'd in ocean's pearly caves;
First forms minute, unseen by spheric glass,
Move on the mud, or pierce the watery mass;
These, as successive generations bloom,
New powers acquire and larger limbs assume;
Whence countless groups of vegetation spring,
And breathing realms of fin and feet and wing.

EXTRACT FROM THE TEMPLE OF NATURE,
ERASMUS DARWIN

DANCING FROGS

Luigi Galvani (1737 – 98). In the latter half of the 18th century, Europe's science elite were stimulated by electrical energy yet it was one Italian scientist from Bologna who, using frog's legs and Leyden jars, made leaps and bounds in this field of endeavour. Some way into his radical research, Luigi Galvani jumped to an erroneous electrical conclusion that would send shock waves through the scientific community and ultimately lead to a deeper understanding of electricity and its uses.

Man of the Cloth to Man of Medicine

Very little is known about the early years of this eminent scientist. Born to a goldsmith and his fourth wife in Bologna on 9 September 1737, Luigi Aloysio enjoyed a privileged upbringing. While not aristocracy, the Galvanis were certainly well-known and sufficiently well-heeled to provide their four children with the best education possible. Initially, Luigi showed an interest in the church and at just 15 years of age he joined the illustrious *Oratorio dei Padri Filippini* where he studied theology.

It was during this time, however, that he changed tack. Contemplating taking his religious vows, he made an academic about-face in favour of medicine, and in 1755 he enrolled in the Faculty of Arts at the University of Bologna. Here Galvani received the wisdom of some of the top minds in their chosen fields. He was taught chemistry by Jacopo Beccari, trained in surgery by Giovanni Galli and attended the courses in natural history and in physics by Domenico Galeazzi.

Following his graduation in both medicine and philosophy in 1759, Galvani wasted no time in pursuing a scholarly career, writing a thesis on bones that won him an honorary yet unpaid position as lecturer of surgery at the University of Bologna. This role allowed him to improve his anatomical knowledge and surgical techniques and he quickly gained a reputation as a skilled teacher.

While exploring the ureters of live chickens as part of his research into the genitourinary tract of birds, the avian-obsessed academic courted the daughter of his physics professor, Domenico Galeazzi, and the pair married in 1762. This brought Luigi into the professor's home and here he assisted his father-in-law in further experiments inside his domestic laboratory.

Such intense involvement in scientific research helped Luigi Galvani perfect his operative skills and as his abilities improved so did his status. By the early 1770s he found himself at the very top of the city's medical hierarchy as president of the Academy of Sciences and his University, bringing him considerable wealth and prestige. It would not be long before he would begin his experiments that would bring him considerable fame and a place in history.

Fried Frog's Legs

Throughout the 18th century electricity had become a real focus of scientific study and it was not long before Galvani took an interest. Concentrating on neurophysiology, the Italian began a series of experiments looking into the effects of electricity on nerves and muscles. He equipped his private laboratory with state-of-the-art apparatus; an electrostatic machine for creating sparks, Leyden jars to store static electricity and various kinds of condensers.

Frog's legs connected to an electrical feed. Luigi Galvani discovered that the leg muscles of dead frogs twitched when struck by an electric spark. Galvani's re-animation experiments were mentioned specifically by Mary Shelley in her diaries before she wrote *Frankenstein* (1818).

Equally essential to his experiments were frogs. Easy to obtain and possessing many helpful properties including muscles that continued to contract long after death, an endless number of these long-legged amphibians croaked in the name of science. Galvani would often be found outside in his garden with a batch of dissected frogs in violent storms studying the effects of atmospheric electrical emissions. Attaching brass hooks between the spinal cord and an iron railing, he would watch as lightning flashed stimulating the leg muscles into action, causing them to twitch madly.

Seeing the frogs perform this strange staccato dance led Galvani to believe there was an intrinsic connection between the frog's movement and the 'natural' electrical discharge and he upped his research. Leaving his laboratory only to fulfill his other academic commitments, the Bolognese boffin worked furiously to substantiate his ideas.

Animal Electricity

It was during this period of intense investigation that Galvani's most famous experiment took place. Many versions of the tale exist but the basics remain the same. One typical day while researching inside his home laboratory the scientist was carefully preparing yet another amphibian for testing. Taking a metal scalpel, he touched the exposed sciatic nerve of the frog and, to his surprise, saw sparks. Suddenly, the dead frog's legs began to kick violently as if brought back to life.

Having spent years analyzing the relationship between electricity and anatomy, Galvani instantly became convinced he had witnessed a previously undiscovered life force within the muscles of the frog. At this time two sources of electricity were known to exist: the natural kind found in the atmosphere such as lightning and an artificial form created by friction. Now, according to Galvani, there was a third source.

He named this new form 'animal electricity' and was adamant he had discovered a special electrical fluid residing within the nerves which compelled the muscles to move. He also went one step further, suggesting this force was so powerful that if one frog ate another frog then it would become a super version and be able to grow up to 12 feet in length!

Versus Volta

In early 1792, Luigi Galvani published the results of over 10 years of electrophysiological research in his famous essay: *A Commentary on the Effects of Electricity on Muscular Motion*. Heading the thesis were details of his shocking discovery. His report on 'animal electricity' contrasted greatly with current scientific thought. Up to this point physiologists held the Hallerian belief that nerves and muscles were made of highly excitable tissues but electricity had no part to play in their operation.

Unsurprisingly, Galvani's findings became an international sensation. His new-found theory spread far and wide as scientists from all over Italy and Europe repeated his experiments. While the majority of experts hailed his discovery of the animal electricity as revolutionary, one of his colleagues was less convinced, questioning the veracity of his findings.

Alessandro Volta, professor of physics at the University of Pavia, had initially embraced the findings. But after replicating the experiments in his own laboratory he became increasingly critical of the claims made by Galvani in his publication. The man from Pavia suggested his fellow scientist had made a mistake in concluding that the strange movement of the frog's legs was due to an electric fluid existing inside the specimen's nerves. In fact, Volta had an altogether more simple explanation.

Volta believed what Galvani had witnessed was actually the result of electricity generated by two dissimilar metals. According to Alessandro, what had happened was the steel scalpel had mistakenly touched a brass hook holding the frog's leg in place. It was the relationship between these two metals that had produced a spark which had then caused the dead muscles to jerk into action. A major dispute between the schools of Bologna and Pavia broke out as further tests by both sides endeavoured to come to a definitive conclusion. The reserved Galvani was reluctant to get involved in the controversy and kept a low profile while Volta went on to create his artificial electric organ called the Voltaic Pile. This stack of conductors – the first battery - came into being thanks to this heated argument which focused his attention on the conductive powers of dissimilar metals.

No to Napoleon

As the conflict raged, Galvani grew weary. Much of his spark had been extinguished following the death of his beloved wife, Lucia, yet he continued to persevere with his experiments. In 1794 he published his second major work on animal electricity entitled: *A Treatise on the Conducting Arc*, which he did anonymously. A year later he made a rare trip outside Bologna to the Adriatic coast where he caught some electric fish. These specimens were used in another series of tests to confirm his theory of internal electric matter and were the focus of his *Memoirs on Animal Electricity*, published in 1797.

With old age setting in, Galvani soon abandoned his research, leaving it to the younger generation to explore the uncharted world of electricity. Besides, he had more pressing issues to deal with. Following Napoleon's invasion of Northern Italy, including the papal city of Bologna, the new Cisalpine regime demanded all university professors to swear allegiance to this new republic. Galvani refused to take the oath believing it conflicted with his religious beliefs. This rebellion saw him removed from all his academic posts and deprived of his livelihood.

Forced to retire to the house of his younger brother, Giacomo, his career and status slumped. Several months later the republican government agreed to reinstate the city's famous scientist but the change came too late. On 4 December 1798 Luigi Galvani died in his home. A marble statue was erected in front of the University of Bologna to commemorate the life and works of this pioneer of electro-physiology, who opened – albeit indirectly – the way for electricity to become the source of power it is today.

RESURRECTION OF THE DEAD

Giovanni Aldini (1762 – 1834). **Popularizing his uncle's concept of galvanism, this Italian impresario showcased a series of shock therapy techniques on a grand tour of Europe in the early years of the 19th century. Striking awe into audiences with dramatic demonstrations on the corpses of freshly-executed criminals, Giovanni Aldini succeeded in persuading witnesses that he could bring the dead back to life, a feat that would later provoke Mary Shelley to pen her horror classic,** *Frankenstein.*

A Family Affair

Born in Bologna on 10 April 1762, Giovanni Aldini was destined to become a man of science. As nephew to Luigi Galvani, the discoverer of galvanism, the course of his career was quickly set by his mother who steered Giovanni's education in the direction of his famous uncle. To this end he attended the University of Bologna receiving a degree in physics in November 1782.

Following his graduation, Aldini went immediately to work as a research assistant in Galvani's home laboratory. Here he helped his uncle in his experiments with frog's legs seeking the apparent electrical fluid believed to reside in the nerves. So adept did Aldini become at this scientific analysis that he helped edit the great man's published works, even adding supplements of his own. It was fast becoming clear that Giovanni was going to become an important scientist in his own right.

Defending Galvanism

Aldini may have been groomed to follow in his uncle's footsteps but he was by no means a carbon copy. While Galvani was discreet and humble befitting his pious nature, the younger man was more of a firebrand bursting with youthful exuberance, keen to excite and quick to blaze. During the quarrel with Alessandro Volta over the existence of animal electricity Aldini took the lead in defending the concept, becoming its most ardent supporter.

Following Volta's insistence that Galvani's frogs moved only because two dissimilar metals came into contact with each other,

Aldini set about proving this was not the case. He acquired some purified mercury and used this as the sole metallic element in his own frog experiment, successfully creating an electric arc and causing muscular contraction.

Volta was quick to respond to this new development. Refusing to budge from his bimetallic theory, the physics professor from Pavia declared Aldini's test succeeded in stimulating the frog's legs not through any internal spark of new-found animal electricity but because the mercury was tainted with traces of at least one other metal.

This led both scientists to begin a series of experiments inducing muscle contraction in frogs without any metals involved. Aldini managed it soon after, producing a muscular spasm each time the muscles were put in contact with the nerves in the leg. Describing this test in the supplement to his uncle's *Treatise on the Conducting Arc* in 1794, Aldini had made a major contribution to the advancement of galvanism.

A Shock To The System

In 1798 Giovanni Aldini became professor of experimental physics at the University of Bologna, taking the chair from his old tutor, Sebastiano Canterzani. It was also the year in which his uncle passed away leaving him the sole torchbearer for galvanism. Despite his growing academic responsibilities the Italian wasted no time in searching for new practical uses for the electrical phenomenon.

His laboratory quickly resembled that of an abattoir as the brains of birds, lambs, calves and oxen became the subject of a series of experiments. Aldini would apply an electrical charge to different sections of the brain to determine their sensitivity. On discovering that he could produce powerful responses from the cerebellum and corpus callosum, he began to consider galvanism as a therapeutic tool.

Trials on human brains produced similar results so he planned to take the next step: apply direct current to a living human head. Not unwilling to put himself at risk in the name of science, brave Aldini volunteered his own for the experiment. Experiencing a vigorous jolt against the inner surface of his skull, he discovered the effect was exaggerated as he moved the electric arcs from one ear to the other. Aside from suffering from insomnia for several days afterwards, Aldini reported no adverse reaction to the procedure and quickly inquired after suitable subjects on whom to repeat the treatment.

The most documented account concerns one Luigi Lanzarini, an Italian farmer in his late 20s who had been committed to Santo Orsola Hospital in Bologna in May 1801. Diagnosed with melancholy madness, the contadino agreed to undergo the ground-breaking therapy and over a period of several weeks Aldini allowed a weak electrical charge to be regularly applied to the patient's shaved head. This primitive course of electro-convulsive treatment led not only to the complete cure of Lanzarini's depression but also to many other successes as Aldini heralded the procedure as a proven remedy for a range of mental illnesses.

Criminal Convulsions

Such achievements in its constructive merits persuaded the irascible Italian to take a more aggressive stance in promoting galvanism. A natural showman, Aldini organized scientific spectacles in his native city to create support for the phenomenon, publicly electrifying the bodies of animals before awestruck crowds. In one particular act he decapitated a dog then attached leads to the head and by way of an electrical charge made the jaw open, the teeth chatter and the eyes roll about in their sockets.

In the early months of 1802 Aldini upgraded to human subjects. Drawing blood-lusty crowds to the large public area before Bologna's Palace of Justice, he gave a number of performances using the freshly-executed corpses of criminals. Employing the recent invention of his fiercest enemy – Volta's bimetallic pile – to provide the electric current, Aldini managed to create the most horrific contortions from the decapitated heads as muscles and nerves flexed and twitched in their faces. Spectators were rendered speechless as the Bolognese bewilderer caused the headless cadavers to move some three hours after death, exciting the vital forces in one felon's arm to miraculously grasp a coin and toss it across the room.

These macabre exhibitions of electrical stimulation quickly attracted the attention of scientific organizations outside Italy. In the autumn of 1802 Aldini visited Paris where he lectured and experimented before members of the Royal Academy of Sciences. He also called in at La Salpêtrière Hospital where he astounded the resident psychiatrist, Philippe Pinel, by seemingly resurrecting an old woman who had recently died of typhus.

Resurrecting George Forster

From here Aldini moved to England; the final destination on his European Grand Tour. His dramatic demonstrations played to packed amphitheatres at Guy's and St

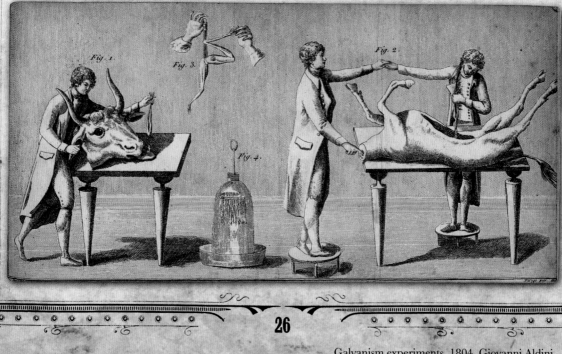

Galvanism experiments, 1804. Giovanni Aldini organized spectacular public performances electrifying dead animal carcasses to prove the theory of animal electricity.

Thomas Hospitals. Physicians, ambassadors, dukes and lords came to witness the gruesome entertainment. Yet it was one particular show that has since gone down in history as Aldini's masterwork.

His *pièce de résistance* took place on 18 January 1803. In the early hours of that Monday morning, the limp body of George Forster was cut down from Newgate gallows and spirited away to a nearby house. According to judge and jury, Forster had drowned his wife and child in Paddington Canal and had been sentenced to hang until dead. Now, his corpse had become the property of the Royal College of Surgeons and was about to be subjected to some terrifying experiments.

Here, before an expectant audience, Aldini prepared to reveal the superior powers of galvanism. Conducting rods connected to a large battery were placed against various parts of Forster's body with dramatic results. When attached to the mouth and ear the spectators watched the jaw quiver feverishly and the left eye wink open. Next came the climax of the show. Aldini took one of the rods and inserted it into the dead man's rectum. This invasive act caused the greatest reaction, causing the legs to kick furiously and a fist to clench tight and visibly punch the air.

This final display left the entire room in a state of shock. Many eyewitnesses returned home believing they had seen the murderer brought back to life. In fact, the performance evoked such horror that Mister Pass, the beadle of the Surgeons' College, died of fright soon after. While Aldini may not have succeeded in resurrecting a dead corpse, what he had achieved was to highlight the potential power of galvanism, to demonstrate its potential efficacy in cases of drowning and asphyxia. For this he was awarded the Copley Medal by London's Royal Society.

The Fright Knight

Following the completion of his tour Aldini documented his electrical experiments in his 1804 work: *Essai Théorique et Expérimental sur le Galvanisme*. He then spent the remainder of his career working on a series of innovations that had more to do with physics than medicine. First he looked at lighthouses, improving their illumination by gas. Then in 1827 he invented a protective garment made of asbestos for firemen, which he tested around Europe. However, Aldini would forever be remembered for his work promoting galvanism.

Much like his Uncle Luigi, Aldini's electrifying demonstrations helped contribute to our understanding of physiology and electricity. His work paved the way for development in various forms of electrotherapy that would become extremely popular during the 19th century. His frightful re-animation of Forster also had cultural ramifications, inspiring Mary Shelley to write her Gothic horror novel, *Frankenstein*.

In recognition of his accomplishments, the emperor of Austria made him a knight of the Napoleonic Order of the Iron Crown and, in 1807, a councillor of state at Milan. It was here that the great Aldini died on 17 January 1834, leaving a considerable sum to found a school for natural science in Bologna.

ELECTRIC MESSIAH

Nikola Tesla (1856 – 1943). Never fully recognized in his own time, this giant of science and electricity patented more than 700 inventions during his life. Designed in the mind via labyrinthine visions, this obsessive-compulsive with a penchant for power developed the technology that permitted the long-distance distribution of electricity leading to a whole new world of scientific possibilities.

A True Visionary

On 10 July 1856 in the rural village of Smiljan deep in the Lika mountains of modern-day Croatia, a Serbian priest and his wife gave birth to their fourth child. His name was Nikola Tesla. Both in and out of school little Nikola proved to be something of a child genius. Not only did he manage to complete four years of education in just three, but he also spent his leisure time immersed in scientific endeavours that would have baffled those more senior in years.

Fascinated by mechanical engineering, one of his earliest inventions was a rotary engine imaginatively powered by insects he had glued to a paper wheel. His mind was filled with such creative visions. In fact, he possessed a photographic memory so powerful he was able to visualize complex designs in precise detail. Despite his love of science, his father was adamant he should enter the priesthood. However, following a bout of cholera that dangerously threatened his life, Nikola's parents allowed him to follow his heart.

And so in 1875 the teenage Tesla headed west to study electrical engineering at the Technical College in Graz, Austria. Here he tinkered with alternating current and witnessed the Gramme dynamo operate as a generator and, when reversed, an electric motor. This got him thinking. His vision-packed mind pondered on how the power of AC could best be utilized. As yet, there was no safe and efficient means to employ this type of current. But perhaps he could develop a way. This puzzle would plague his thoughts over the coming years.

A Spark at Sunset

By the age of 24, Nikola had had enough of established education. Having a brief spell at the Karl-Ferdinand University in Prague, he moved to the prosperous Hungarian capital of Budapest to look for work. He thought he'd found his calling in the national telephone company, working his way up to become chief electrician of the country's new exchange. However, his

Nikola Tesla, inventor and electrical giant.

thoughts were elsewhere. The conundrum of alternating current had become an obsession and coupled with his demanding day job his mind burned out.

Tesla suffered a complete mental and physical collapse. An oversensitivity to light and sound, while believing the very air around him was on fire, defied medical diagnosis. Eventually he returned to normal. Taking gentle walks in the open air seemed to stabilize his condition. And it was during one such walk that the recuperating Nikola came upon his greatest discovery.

One February evening while taking a sunset stroll in the local park with a co-worker he suddenly froze, his marvellous mind gripped by a brainwave. After years of mulling over his AC power problem, a perfect picture of the solution had finally come to him. With his friend looking on, Nikola feverishly sketched the design in the dirt. Two months later he had transplanted these makeshift doodles into intricate blueprints for motors, dynamos and transformers. Now he just needed the money to manufacture these plans.

Nikola Tesla in seinem Laboratorium in den Colorado=Bergen unter künstlichen Blitzen.
Nach einer Professor Slaby zugeeigneten Photographie. (Aus A. Slaby, Glückliche Stunden. Berlin, L. Simion Nachf.)

Nikola Tesla nonchalantly reads a book in his Colorado Springs laboratory in 1899, while the giant Tesla coils create artificial lightning all around him.

Testing Times for Tesla

A few months following this revelatory vision, the well-again Tesla moved to Paris, joining the Continental Edison Company. While on assignment in Strasbourg he made the next step towards making his power plans a reality. The German city had invested in Edison's direct current system but a dangerous explosion at the dedication ceremony prompted the owner to send Tesla to perform a year's worth of repairs. It was during this time that he rented a machine workshop and built his first AC induction motor.

Unable to acquire financial support back in Paris, Nikola crossed the Atlantic to New York in the summer of 1884 with very little to his name in the hope of finding a job as well as patronage for his AC power. He found the former with an engineering role in Edison's American company but the latter proved impossible to obtain owing to the dominance of direct current.

Tesla soon left Edison's employ under a cloud after the owner reneged on a deal that would have netted the Serbian a cool US$50,000. Another setback followed in 1886. After finally managing to set up his own company, Tesla Electric Light and Manufacturing, his financiers chose to pull out leaving him destitute. The expert engineer was then forced to dig ditches to keep from starving.

Over the next 18 months Tesla managed to scratch together sufficient funds to construct an advanced induction motor which he demonstrated before the American Institute of Electrical Engineers on 16 May 1888. The device worked perfectly and was lauded as a triumph of electrical science.

News of his achievements spread and soon came to the attention of one man who was prepared to listen to his ideas.

The War of Currents

George Westinghouse had made his millions in the rail air-brake business and he quickly saw Tesla's potential. With his commercial acumen and Tesla's technology together they began to develop alternating current systems around the country. A lucrative contract between the pair allowed the inventor to focus on other creations conjured up by his brilliant mind. Soon he was touring throughout the States and Europe shocking audiences with his wireless energy transmission, creating great discharges of electrical fire with his Tesla Coil.

Meanwhile, Edison had embarked on a propaganda campaign to discredit the new current. False claims of instability and danger to potential clients were made by Tesla's former employer. He went as far as to electrocute stray cats and dogs to demonstrate the high risk. The truth was somewhat different. In fact, alternating current was more efficient than Edison's system. Unlike the prevailing power, it did not require expensive generators every two miles to distribute to users.

In May 1893 Westinghouse illuminated the World's Columbian Exposition in Chicago with AC current. Tesla also performed his astonishing demonstrations to visitors, allowing a charge of one million volts to pass through his body without harm, thus disproving Edison's allegations. These successes helped to win the contract to install the first powerhouse at Niagara Falls. In

August 1895 the system went online and began carrying power to Buffalo over 20 miles away. The war of currents was now won.

Frightening Lightning

Not one to rest on his laurels, Tesla allowed his fertile mind to form new inventions and soon his ideas outgrew his New York lab. In May 1899 he moved to Colorado Springs, Colorado where he acquired the perfect workspace. With its roll-back roof it could cope with the high-voltage tests he had planned. Using his giant Tesla coils – over 75 feet in diameter – and a magnifying transmitter he was able to create artificial lightning more powerful than nature. Electrical emissions 135 feet long emanated from the laboratory, the accompanying thunder was heard over 15 miles away.

These extreme experiments caused considerable disruption to the city. Blackouts were common, so too were flames of electricity shooting from household taps. Unsurprisingly, Tesla left Colorado six months later.

His next project was his biggest yet. With funds from J. Pierpont Morgan, he set about establishing a transatlantic wireless communications facility on Long Island, called Wardenclyffe. A 154-foot-high domed tower was designed to send power and information across the world. Sadly, the project failed. Morgan pulled his funding and Nikola suffered another emotional breakdown.

His mental state manifested itself through manic compulsion. An obsession with the number three caused him to circle a block three times before entering a building and to stay in hotel rooms whose numbers were divisible by three. He harboured a revulsion for jewellery and formed a pathological fear of contamination. While such phobias certainly hurt his reputation, they did not restrict his ability to create.

The Death Ray

Living off a modest pension from friends in Eastern Europe, Nikola continued to invent inside his rented room at the Waldorf-Astoria Hotel. In 1934 he announced to the world's press that he had notionally discovered what he called a teleforce weapon.

Electrostatically-charged particles passed through a vacuum tube to create a thread-thin stream of atomic clusters which, he believed, would create a ray so powerful it could destroy a fleet of 10,000 aircraft at a distance of some 200 miles. While Nikola saw it as more of a defensive device, acting as a protective force field, reporters were quick to point out its obvious offensive capabilities, christening it the death ray.

Tesla's superweapon attracted little interest at the time and the aging ace of alternating current lived out the remainder of his days in a two-room suite at the Hotel New Yorker. Following his death on the 7 January 1943, he was given a state funeral, attended by over 2,000 people. Past Nobel Prize winners acted as pallbearers and his ashes were later sent to Belgrade. Crates filled with his laboratory notes were also seized by the FBI and declared top secret. To this day, scientists still pore over the contents hoping to uncover something new.

RE-ANIMATING A KILLER

Andrew Ure (1778 – 1857). **Eight months following the release of Mary Shelley's** *Frankenstein*, **life imitated art when a Glaswegian galvanist turned Gothic fantasy into morbid reality in a university amphitheatre. An accomplished astronomer, geologist and chemist, Andrew Ure is remembered less for his astounding achievements in these fields and more for one November day in Glasgow when he demonstrated the possibility of re-animation.**

Ure's Lectures

Little is known about the childhood of Andrew Ure. Born in Glasgow on 18 May 1778, the son of a cheese-monger, his early education must have been sufficiently erudite to have won him the opportunity to attend university. Studying at both Edinburgh and Glasgow, he achieved his MA by 1799 with a notable essay on hernia, winning the university prize for anatomy. Clearly having caught the medical bug, Andrew went on to receive his doctorate, graduating in 1801. Following a two-year stint as an army surgeon with the 7th Regiment of the North British Militia, he eventually settled in his native city as a private medical practitioner, and was elected to the Faculty of Physicians and Surgeons in 1803. A year later saw Doctor Ure make a return to academia, taking the chair of natural philosophy and chemistry at the newly-established Andersonian Institution. Respected for his enterprising research, the professor ran a first-class laboratory, considered to have been the first in Europe to have held classes in practical chemistry.

Often regarded as self-righteous and combative by his colleagues, this aggressively ambitious academian was highly revered by those outside his scientific circle. This had much to do with the popular public lectures he gave at the Andersonian. Every Tuesday and Friday evening Ure enlightened the attendees on an array of disciplines, displaying his encyclopedic knowledge of mechanics, astronomy, pneumatics, hydrodynamics and more. These lectures were so successful that they inspired the creation of new scientific societies throughout Britain and even into Europe.

Over the ensuing years Ure's growing reputation endowed him with honours and accolades from institutions all over the world. His wide-ranging abilities saw him working in a number of unconnected fields. He designed and oversaw the installation of a 14-foot telescope during his residency at London's Garnet Hill observatory. Three years later, Ure was in Belfast consulting for the Irish linen board, for whom he developed techniques to analyse acidic and alkaline content. Yet these achievements would be comprehensively overshadowed by one particular experiment back in his native Glasgow.

A Gander at the Gallows

On 4 November 1818 there was a palpable air of apprehension in the streets of Glasgow. It had been nearly 10 years since the last execution in the city and the public were restless, eager for their bloodlust to be sated. Throngs of thousands had poured into Jail Square at the bottom of Saltmarket, pushing and clambering for a view of the gallows built before the new High Court

building. The authorities were forced to post guards at the timber footbridge across the River Clyde, fearing the crowds eager to gain a clear vantage point would overload the structure and cause its collapse.

The star of this morbid show was one Matthew Clydesdale. Two months earlier this Airdrie weaver had been arrested for the murder of a 70-year-old man while in a drunken rage. Brought before the Glasgow Assizes on 3 October, Clydesdale was found guilty by Lords Succoth and Gillies and sentenced to death by hanging. Sometime after 2 p.m. on that November day, the muscular murderer was escorted to the gallows where a blindfold covered his eyes and a rope placed about his neck. Minutes later, hangman Tammas Young dropped the convict for the *coup de grâce* bringing cheers from the buzzing crowd.

Last Call for the Condemned

Yet this was not the end of proceedings. Clydesdale's sentence had a second clause:

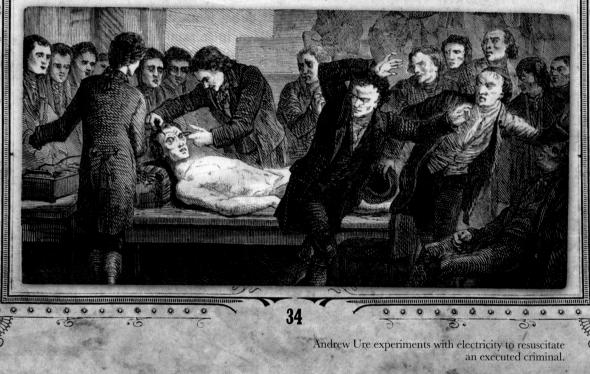

Andrew Ure experiments with electricity to resuscitate an executed criminal.

his body was to be submitted for anatomization. To this end, the dead offender was cut down after an hour of hanging, placed on a dray cart and transported post-haste to Glasgow University. Many inveterate voyeurs awaited his arrival, crammed cheek by jowl inside the packed anatomy theatre to await another spectacle.

The man behind this additional performance was none other than Andrew Ure. Using his innate assertiveness, the enterprising doctor succeeded in persuading the authorities to permit him the opportunity to experiment upon the corpse of Matthew Clydesdale. No sooner had the convict been condemned than Ure began pressing for the remains. And somehow, despite having fallen out with a number of high-ranking officials over the years, his request was granted. But there were conditions.

The University hierarchy insisted the event should follow official procedure; running the show would be their Professor of Anatomy, Botany and Midwifery, James Jeffray. Also, all dissection was to be undertaken by Thomas Marshall. These provisos and posturing were nearly rendered pointless when Clydesdale slashed his own throat with a broken beer bottle. Much to Ure's relief, the prisoner survived the suicide attempt to undergo a series of bizarre experiments.

How To Make A Point

As the corpse made its inexorable journey to the amphitheatre Andrew Ure made last-minute adjustments to his principal instrument; inside a mahogany case was a device of his own design, a huge galvanic battery. Consisting of 270 pairs of zinc and copper discs, each 4 inches in diameter, this electric pile had to be charged with dilute nitric and sulphuric acids in preparation for four distinct demonstrations the doctor had planned.

The first test required an incision made at the neck, through which a section of vertebra was removed, exposing the spinal cord. Two successive cuts were then made in the left hip and the left heel. Next, Ure took cylindrical rods connected to the battery and attached them to the gashes at the neck and hip. Suddenly, the body of Matthew Clydesdale seemed to come alive, shuddering violently before awestruck spectators. The second rod was then moved from hip to heel inducing a knee-jerk reaction in the corpse. The left leg suddenly straightened, kicking out and almost knocking over an assistant.

The second experiment involved stimulating the respiratory system. By placing the rods to the phrenic nerve and diaphragm then providing intermittent electrical charges, Ure was able to cause the chest to rise and fall. To those present it seemed as if this dead man was now breathing.

The third demonstration was perhaps the most dramatic. Using the charged rods once again, Ure excited the supra-orbital nerve in the forehead, applying 50 shocks in just two seconds. Variations in the voltage caused the felon's face to run a whole gamut of expressions from rage to despair. These macabre looks spooked a number of witnesses into leaving the room, one gentleman reportedly fainted.

The final test was equally chilling, causing the hair to stand up on the necks of the onlookers. A simple cut was made in

Clydesdale's forefinger and the rod placed inside. Activating the battery then sent a violent charge into the digit forcing the executed convict to eerily raise his hand. So strong was the force that the hand resisted all attempts to hold it down and even began pointing to various members of the audience.

A Real Rusurrection

Despite receiving such public attention, only one of three city papers bothered to cover the shocking story. Andrew Ure and his resurrection techniques seemed destined to be forgotten in time. However, almost 50 years later a Glaswegian writer named Peter Mackenzie performed his own resurrection, reviving the tale of the doctor's demonstrations. He claimed to have been present in the theatre that day, declaring that Clydesdale actually came back to life before an alert anatomist grabbed a scalpel and slit the re-animated corpse's throat.

Mackenzie was clearly spinning a yarn. Rather than intent on restoring the life of the recently dead, Ure was more interested in the galvanic properties of the electrical charge sent through the cadaver. That's not to say he rejected all forms of resurrection. In fact, Ure suggested afterwards that if the battery was connected to a subject pronounced dead, in some cases resuscitation would be possible, for instance after asphyxiation or drowning. This proposition effectively described the modern day defibrillator used around the world to save lives.

What a Load of Rubbish

During these ghastly tests, the Glaswegian was in the middle of a bitter divorce trial. He had accused his wife of cheating with a Granville Pattison, one of his fellows at the Andersonian Institution. By the following year the divorce was finalized and Ure was free to dedicate his time to a variety of scientific pursuits. In 1821 he published his first major work, *A Dictionary of Chemistry* which led to his election to the Royal Society. He followed this début with a number of publications covering geology and manufacturing.

As well as a prolific writer, Ure excelled in more practical endeavours. His versatile mind was in high demand for a variety of industries and as a consulting chemist was asked to perform an array of chemical researches. This included analysis of coal for the Admiralty, bread-baking processes and tests on sugar beet and sugar cane. He even solved the mysterious Pimlico Malaria tragedy. Sewage workers were perishing below ground and commissioned tests declared it was due to mosquitoes. Ure's independent study debunked this outrageous notion, declaring the deaths were down to poisonous fumes emanating from dumped rubbish.

As old age set in, Ure suffered repeatedly from an acute case of gout, which for many years attacked his right side. In retirement, he travelled throughout Europe with his devoted daughter, Catherine, seeking treatment for the affliction at a number of spas. Yet while his body was failing him, the erudite chemist remained mentally as sharp as a tack until his dying day. This came on 2 January 1857 in the London home of his son, Alexander.

ELECTROTHERAPY PIONEER

Christian Gottlieb Kratzenstein (1723–95). From measuring steam particles and synthesizing sound to designing airships and arithmetic machines, this prize-winning polymath spent his entire career innovating and inventing in a number of scientific fields. Yet it was his work with electricity that secured Christian Gottlieb Kratzenstein his place in history becoming one of the first pioneers in electrotherapy; discovering the therapeutic potential of static power.

Electrifying Education

Born on 2 February 1723 in the small German town of Wernigerode, Christian Gottlieb grew up in the shadows of the rugged Harz Mountains, his head filled with electric dreams. As with many in the 18th century, electricity had captured his imagination and, from an early age, this second youngest son of a high school teacher demonstrated a deep curiosity in the physical phenomenon. While working his way through school he would perform experiments with electrical friction machines and regularly frequent the Brocken – the highest peak of the region – to marvel at the extreme weather conditions it endured.

Following his time at the lyceum, young Christian attended the University of Halle, choosing predictably to study physics and natural sciences. Here he came under the tutelage of the new professor of philosophy and medicine, Johann Gottlob Krüger, which was to have a real impact on his future. During a series of lectures, Krüger suggested the electrostatic or friction machines fashionable among scientists may have a place in medicine. Believing all things should have a usefulness, the professor proposed that electricity was no different, and could possibly have a beneficial effect on paralyzed limbs. Inspired by his tutor's hypothesis, Kratzenstein set out to test his ideas.

Therapeutic Thrills

Kratzenstein immersed himself in the pioneering work of past scientists. The research of Otto von Guericke, Francis Hawksbee and Stephen Gray helped guide his own investigation into the possibilities of electrical stimulation. Using human subjects, the energized student found that a charge could increase the heart rate by at least 10 per cent, improve the circulation and cause the muscles to contract. These initial findings led him to speculate such electrification could help a wide range of afflictions.

He began bringing patients to his home, sitting them upon a wooden stool and using Leyden Jars and electrostatic generators with their large revolving frictional glass globes to draw sparks from the affected body parts. Kratzenstein would even apply the same process to himself and soon discovered regular doses of this primitive electrotherapy throughout the day would improve the quality of his sleep at night. As well as a potential cure for insomniacs, he predicted the procedure could help with other mental ailments including hypochondria and female hysteria!

Yet it was its remarkable effect on physical disorders which attracted the greatest attention. One pertinent experiment undertaken by Kratzenstein was on a male and female patient, both of whom suffered from arthritis. The woman complained of a painful contracture in her small finger however, after a 15- minute burst of electrostatics, she was able move the tiny digit without discomfort. As for the man, a similar electrical spell rendered his lame fingers sufficiently movable to play his beloved harpsichord again. These test subjects were possibly the first people to benefit from electrotherapy. Kratzenstein, still only 21 years of age, published the details of his research in his thesis: *Theoria electricitatis more geometrico explicata.* Word quickly spread regarding these new

therapeutic advantages of electricity. Suddenly this physical phenomenon appeared to possess the powers to alleviate both physical pain and mental anguish.

Blowing Off Steam

His studies did not end with electricity. In 1743 he entered a contest set by the Academy of Science in Bordeaux which offered a prize for the best paper concerning the theory of steam elevation. Once again, he plunged himself headlong into research. He constructed a cloud chamber, examining the steam rising from the surface of hot water with a magnifying glass. He then concluded, wrongly, however, that the steam particles were invisible air-filled bubbles, and estimated the diameter of the diaphanous globes to be 1/12th the size of a human hair. His resultant essay on evaporation won joint first place.

By the time Kratzenstein acquired his Master's degree and doctorate in 1746, his name was already on the lips of the scientific elite. His developments in electrotherapy had gained considerable fame and his talents were highly sought after throughout Europe. Despite this, the gifted scholar chose to remain at Halle, becoming part of the faculty, lecturing in medicine and physics. Proving to be a first-class teacher he was loved by his students, yet his aggressive manner often led to clashes with his colleagues.

A Role in Russia

Despite his lack of grace and good nature the hot-headed academic from the Harz

Mountains had enough friends in high places to be granted entry into the Leopoldina. This was the prestigious German Academy of Sciences named after Holy Roman Emperor Leopold I and held to be the world's oldest learned society. However, Kratzenstein received this honour just as he was about to leave his native country for pastures new.

Seeking a new challenge, he answered the call from the Russian Academy of Sciences in St Petersburg and in the summer of 1748 became their professor of mathematics and mechanics. For the next five years he worked on a whole range of tests and innovations. A true polyhistor, he covered all fields of science developing astronomical instruments, inventing nautical weights and

designed a rowing aid for ships when cursed by calm waters. Even when the antique Globe of Gottorf, a water-powered orb of the Earth that demonstrated the movement of the stars and planets, was damaged by fire Kratzenstein was able to complete its reconstruction. It seemed nothing was beyond the abilities of this German genius.

In 1752, while on a fact-finding expedition to Archangelsk remapping the Norwegian coast, he stopped over in Copenhagen and was promptly offered a chair at the university. Keen to escape the political atmosphere in Russia where foreigners were regarded less favourably than when he arrived, he decided to leave St Petersburg in August 1753.

Fire Hazard

Swapping Russia for Denmark, Kratzenstein revelled in his position as professor of experimental physics. Befitting his broad knowledge, he lectured up to six hours a day on all disciplines and his physical demonstrations were lauded by his students. Regrettably, the ordinary citizens of Copenhagen were less enamoured with the university's new addition. Distrustful of his tests with electricity, landlords refused to house the scientist who would surely induce lightning and create fires with his experiments in electrotherapy.

Such concerns were not without foundation. Days before Kratzenstein's arrival in Denmark a fellow German physicist by the name of Georg Richmann was electrocuted to death by a bolt of ball lightning. What turned out to be his final experiment in atmospheric electricity blew open his shoes

Newly Improved Magneto-Electric Machine. A very rare, pocket-sized electrotherapy box from the 19th century.

and destroyed the door of his room.

After eventually being granted lodgings, the foreign fire risk was instructed by the King of Denmark to increase his studies in electricity and observe its effects on the sick. To this end, Kratzenstein issued a proclamation to the people of Copenhagen. Inviting the unwell to his home three days a week, he endeavoured to cure his callers from an array of illnesses by way of electricity. Using a large apparatus with spinning glass globes Kratzenstein created sparks and an overpowering odour from his patients. This smell he deemed to be sulphur being driven out by the electricity leaving the blood cleansed and the body healed.

Prized Possessions

The royal command did not diminish Kratzenstein's other scientific interests during his tenure. In 1758 he completed an encyclopedic work on experimental physics that quickly became essential reading for students over the next 50 years. He also invented a calculating machine superior to the Leibnitz device created a century earlier. Presented before the Imperial Academy in St Petersburg, the Russians marvelled at its operation and wished to obtain the arithmetic instrument. Its inventor, however, refused to give it up poetically stating: *I do not want to send this daughter of my thoughts into exile, since she is unique and I do not possess anything that could replace her.*

Kratzenstein did not always hide away his designs and over the coming years they came thick and fast. To aid his geomagnetic activities on trips to Trondheim during the 1760s he developed his own declinometer; a device used to calculate the angle between magnetic and true north. Then there were his renowned endeavours in the field of phonetics.

In 1779 the annual competition set by the Academy of Sciences in St Petersburg focused on speech. It called for methods to determine the nature and character of vowel sounds, and also a workable instrument to accurately express them. Kratzenstein worked with organ builder Franz Kirschnick to create a series of acoustic resonators, one for each of the five vowels. Inspired by the Chinese mouth organ, or sheng, it worked perfectly before the Imperial judges, winning the prize. It also contributed to the understanding of speech in physiological terms.

Blazing Genius

Even into his 60s his genius showed no signs of abating. He conceived one of the first airship designs, preceding their military use by nearly 200 years. In the same year he identified the cause of the dry fog that had swept over Europe. Kratzenstein deduced the deadly haze was sulphur dioxide resulting from the eruption of Iceland's Laki volcanic fissure.

In 1786, after four decades at the university, he retired from academia. In the summer of 1795 the great fire of Copenhagen struck the city. For two days in June the conflagration razed large parts of the city to the ground, including Kratzenstein's home. His vast collection of manuscripts and instruments was lost to the world and sadly a month later so was he.

ELECTRIC SLEEPING RABBITS

Louise Robinovitch (1881 – 1942). Crossing the Atlantic with a plethora of academic awards and degrees, this shrinking shock-doc wowed the West with her 'exciting' experiments. Zapping rabbits with electric current, the reclusive Robinovitch demonstrated the numbing properties of this electric sleep and went one better; bringing her cotton-tailed test subjects back to life with a volley of volts, proving her years of research were anything but harebrained.

Egghead Emigrates

Scant details survive concerning the early years of this Russian experimentress. Born in 1881 in the Black Sea port of Odessa as Luisa Rabinowitch she was the only daughter of four children. During her early education she exhibited a highly-developed intellect; while the other children played, a withdrawn Luisa studied, and she soon became something of a child genius. Consequently, when her family chose to emigrate from the motherland to America it was decided their *wunderkind* would move to France in order to fulfil her academic potential.

Luisa did just that. Aged just 15, she was awarded a *bachelier ès lettres* (equivalent to a Bachelor of Arts degree) from the Sorbonne in Paris. Following her graduation she then made the trip across the Atlantic where she continued her studies at the Women's Medical College of Pennsylvania. With a handful of honours and a line of letters after her name, she arrived in New York in 1891, settling with her brother Joseph in an East side Russian community.

Anglicizing her name to Louise Robinovitch, she gained employment at Blackwell's Island lunatic asylum as an assistant physician. Here she witnessed first-hand the primitive treatment given to inmates. The uncooperative and out-of-control were chemically restrained with large doses of hyoscyamine, producing partial paralysis. Disgusted by these shameful abuses, the usually-reserved Robinovitch publicly exposed these acts calling for all of New York's institutions to be brought under State control.

Electric Sleep

Disliking the attention her outburst brought her, the émigré doctor retired to Nantes in

ELECTRIC SLEEPING RABBITS

France to focus on her studies away from prying eyes. Mixing with a number of eminent French medical men including Stephane Leduc, Louise completed her thesis at the École de Médecine earning her yet another degree in 1907. It covered in detail her series of experiments in electrically-stimulated anaesthesia or 'electric sleep'.

After years inside a laboratory devoting the best years of her life to the study of electricity on live cats, dogs and rabbits, Dr Robinovitch believed she was close to a breakthrough discovery. Continuing the work begun by her colleague, Professor Leduc, she found that by applying a weak current of around six volts to a live specimen she could place the animal into a conscious state of paralysis. This allowed her to pinch, prick and even operate without causing pain to the subject.

Such a revelation had far-reaching possibilities. If it was possible to render an animal senseless to pain in this way then the same procedure could be applied to humans. At a time when chloroform, ether, morphine and even cocaine were used to knock out a patient at great risk of death, Dr Robinovitch's electrical anaesthetic was a godsend.

Robinovitch Revivals

Following a series of private demonstrations revealing the pioneering benefits of electric sleep the female physician pushed on with her research. She began to test the effects of stronger currents on animals, keen to discover what advantages a higher voltage would bring. After countless failures Dr Robinovitch managed to perfect a technique whereby she could use rapid pulses of electrical stimuli to restart the heart and respiration.

Ready to unveil her findings, the diffident doctor presented a paper before the French Biological Society in Paris on 22 February 1908. It caused a sensation and soon after she was given laboratory space at the Sainte-Anne Hospital in Paris and encouraged by its chief physician, Dr Valentin Magnan, to work on her new method of resuscitation. Months later, before a host of prominent physicians, Dr Louise disclosed her improvements in the manipulation of electricity.

She placed a rabbit under the influence of electric sleep and performed a slight operation, revealing the numbing effects of the procedure. But this was just a foretaste of what was to come. Robinovitch then electrocuted the animal and proceeded to bring it back to life with a series of controlled zaps. The stunned guests were then invited to a first-floor amphitheatre where they were introduced to one of the doctor's patients.

The invalid had apparently lost all power and sensation in his right side. And yet Louise declared that in less than a month she had cured him through a series of electrical stimulation. The once semi-paralysed man informed them that he had exhibited no adverse reactions to the procedure and even enjoyed the sensation. This revelation brought further applause.

The doctor from Odessa seemed to have the magic touch, bringing her fame throughout France. Around this time one of the many morphine eaters of Paris was admitted to hospital and fell into a coma while awaiting examination. All attempts by the physicians to revive the addict failed and so after 20 minutes Dr Robinovitch's equipment was brought in. Electrodes were attached to the lifeless body and the power engaged. In next to no time the apparently dead woman opened her eyes, fully roused from her deathly syncope. This case of resurrection seemed to suggest electricity was truly a life-restoring force.

Edison's Energizer Bunny

Word of Robinovitch's achievements quickly travelled across the Atlantic and various establishments picked up on the potential benefits. The Edison Company, who had suffered a large number of electrocuted employees in their powerhouses, were keen to learn more about this form of resuscitation and so invited Robinovitch to demonstrate her technique. The media-dodging doctor agreed to perform on one condition: that reporters were excluded from the laboratory.

Promises were made and on Thursday 18 November 1909, she gave a demonstration of electrical anaesthesia and resuscitation on a rabbit. The 40 select guests of the New York Edison Company watched with wide eyes as the oblivious bunny had needles inserted through its legs following a mild application of current. The voltage was then increased to a fatal dose, killing the cotton-tailed creature. Leads were attached to its spine and a series of rhythmic excitations caused the heart to restart. Minutes later, the rabbit was hopping about the room much to the observers' astonishment.

Despite the media ban, news of these tests leaked to the press and soon the papers were extolling the talents of this strange little Russian. Articles abounded on her life-rekindling experiments and her refusal to talk to journalists only helped to paint her as an archetypal mad scientist: reclusive, enigmatic and fixated on her work.

Cold Feet

As embellished tales of her work continued to sell papers, Robinovitch spent more time honing the techniques of electric sleep. On 25 January 1910 she helped perform an operation at St Francis's Hospital in Hartford, Connecticut. A 23-year-old Austrian man had been admitted with frozen toes that had now turned gangrenous. Four digits required amputation and Louise's innovative anaesthetic process was applied.

The operation lasted 45 minutes and the patient remained fully conscious throughout, experiencing no signs of pain thanks to the numbing properties of the electric current. In fact, the Austrian spent the majority of the time laughing and joking with the attending surgeons while his toes were removed.

Clear successes such as these led the doctor to consider further ways in which her electrical processes could benefit the human race. The night before the amputation operation she had lectured members of the Hartford Medical Society on the possibility of reviving those supposedly killed by electricity. Robinovitch believed her rhythmic excitations could restore the pulses of those who had undergone the life-taking treatment of the electric chair.

Doctor Deception

Fame continued to follow the shy doctor thanks to her breakthrough experiments, but in late 1910 she would hit the papers for all the wrong reasons. In December of that year her younger brother, Joseph Robin, was arrested. A high-flying banker in the city, he had single-handedly stolen US$207,000 from Washington Savings Bank and caused the collapse of another.

Indicted on eight counts of larceny, Joseph relied heavily on the help of his older sister, who signed an affidavit swearing she was his next of kin. But Louise fell foul of the law herself when their immigrant parents showed up in the courtroom. Both siblings denied the elderly pair were their mother and father so

when they revealed letters of proof, the good doctor was promptly indicted for perjury, spending some time at the Manhattan house of detention known as the Tombs.

Little is known of her life following this brush with the law. She sold her house in 1915 and slipped into the shadows; her career in tatters and in desperate need of resuscitation by electricity or any other means. Her name did appear in the odd article whenever the press wished to rehash those experiments that had made her famous, but the shy doctor's work largely went unnoticed and undocumented. Living the rest of her life how she would have wanted, out of the public eye, Dr Louise Robinovitch died in 1942, aged 71.

American Cardiac Surgeon Professor Claude Beck (centre) in the resuscitation operating laboratory of the Medical School, Case Western Reserve University, Cleveland, Ohio. Professor Beck is credited with inventing the heart defibrillator with which we are all so familiar today. In 1947, during an operation Beck was performing on a 14-year-old boy, the patient went into cardiac arrest. Beck applied an electric shock defibrillator directly to the boy's heart and he made a full recovery.

The Mechanics of Facial Expression.

ABOVE Duchenne de Boulogne and an assistant 'faradize' the muscles of a patient's face.

Guillaume Duchenne de Boulogne (1806–75) was a French neurologist who revived Galvani's electrical energy experiments and contributed greatly to the understanding of neurology. He is remembered especially for the way he triggered facial muscle contractions with electrical probes, recording the grotesque expressions with another recent invention, the camera. In 1862, he published his extraordinary photographs in a book entitled *The Mechanism of Human Physiognomy*.

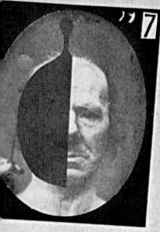

TWO-HEADED FRANKENSTEIN HOUNDS

Vladimir Demikhov (1916 – 98). Deep behind the Iron Curtain inside a top-secret research laboratory, barking mad scientist Vladimir Demikhov took animal testing to a whole new level. Using canines test subjects this Frankenstein physiologist pioneered new techniques in cardiac surgery through a series of macabre experiments in the 1940s and 50s. His double-hearted hounds and dogs with two heads stunned both his peers and the press alike ensuring when it came to extreme science he remained firmly ahead of the pack.

A Call for Discovery

Vladimir Petrovich Demikhov was born into a Russian peasant family on 18 July 1916 at the height of Soviet strife and tension. Just a year later, the October Revolution brought further turmoil to the country as civil war took over. It was during this national unrest that his father would perish, forcing his mother to bring up young Vlad and his two siblings by herself. Labour camps and gulags began to sprout throughout Russia as Stalin's Great Purge saw millions of citizens incarcerated, facing certain death each winter.

During such adverse conditions, Demikhov's mother somehow managed to provide her children with an education and Vladimir studied at a vocational school for working youth where he learnt to become a fitter. Here his interest in physiology was starting to take shape, making a steel copy of a human heart. However, this creation would pale in comparison to future designs by this young Russian.

In 1936, Russian Nobel Prize winner, Ivan P. Pavlov – another scientist known for his experiments with dogs – wrote a famous letter months before his death entitled: *A Letter to Scientific Youth*. It spoke of a need for scientific discovery, calling for young Soviet students to pursue the impossible in defiance of the hostile economic conditions facing Mother Russia, insisting they experiment as far as human ingenuity would allow.

This rousing missive inspired the now 20-year-old Demikhov who had now followed his heart and entered the biology department of the Moscow State University. Here he developed a flair for research and within a year captured the interest of the Russian press thanks to ground-breaking cardiac-related tests on canines. This marked the start of Demikhov's progress in the realm of physiological science.

Repairing the Dogs of War

Graduating from Moscow University in 1940, Demikhov began working as an assistant in its department of human physiology. His progressive investigations led him to consider replacing the heart in one of his lab dogs and was about to undertake the world's first intrathoracic transplantation when World War II interrupted his work. Torn from his studies, Demikhov was called to work on endless mutilated bodies in the overcrowded Red Army hospitals upon the battlefield.

His surgical skills suitably honed, Vladimir resumed his experiments in 1946, performing numerous tests with the hearts and lungs of his canine test subjects. Still riding the wave of free scientific thought, Demikhov believed wholeheartedly that it was possible to transplant major organs in human beings; a concept considered more science fiction than science fact at that time. Upon a backdrop of extreme poverty, repression and state-appointed persecution that saw millions perish on the orders of Stalin, Demikhov set his sights firmly on human survival.

Teaching Old Dogs New Tricks

Now his studies began in earnest. With his experiments not supported by the state, Demikhov was forced to work in secret late at night. Grabbing any moment he could inside his laboratory, feverishly testing new cardiac treatment techniques, he was quickly regarded as a fanatic. He experimented on yet more dogs, connecting and re-connecting blood vessels over 700 times until he had perfected his innovative methods.

These exercises soon paid off when on 30 June 1946 he managed his first successful heart and lung transplant. His canine patient survived nine hours and 26 minutes. A few months later, on 13 October, a similar operation led to the dog surviving five full days. These successes spurred on the Russian and over subsequent years he developed 24 new techniques of heart transplantation. Into the 1950s Demikhov's experiments were becoming more and more advanced, pushing back the boundaries in the name of science. He routinely replaced the hearts of dogs with artificial blood pumps and even implanted additional hearts into their chests, removing part of the lung to make room. He managed to keep this double-hearted dogs alive for as long as 10 weeks.

Demikhov's Frankenstein Dogs

One bitter cold winter night in February 1954 Vladimir Demikhov undertook the experiment for which he would forever be remembered. Donning rubber gloves and gauze masks, the fanatic physiologist and his team of surgeons set to work on a transfor-

mation of Frankenstein proportions. They anaesthetized a large German Shepherd, making an incision at the base of its neck, exposing the jugular vein, aorta and section of the spinal cord. Next, they drilled two holes through one vertebra and threaded two plastic strings through each hole.

After 40 minutes, a second smaller dog was brought into the theatre, its limp form placed next to the larger subject. The team exposed the blood vessels, tying them off individually before severing the spinal column. The next phase was the trickiest. The main blood vessels of the smaller dog had to be connected perfectly with the corresponding vessels of the host dog. Using a surgical stapling machine, a special Russian invention, the pair of pooches were quickly

fused together. The smaller dog's heart and lungs were then cut away and the plastic strings pulled taut to secure what remained to the German Shepherd.

After the marathon operation was complete, Demikhov had succeeded in grafting the head, shoulders and forelegs of one dog to the neck of another; their circulatory and respiratory systems connected to form a two-headed monster, the second head kept alive by the blood pumped by the larger dog's heart. With the two patients' vital signs stable, Demikhov and his assistants waited with anticipation for the anaesthetic to wear off. How would the dog heads react? How conscious would they be? They did not have to wait long for these answers.

Vladimir Demikhov's two-headed dog. The head, neck, shoulders and forelegs of a small dog were grafted onto the neck of a bigger dog in Demikhov's organ transplant lab.

Alive and Licking

As the two-headed beast slowly woke from its deep sleep, the surgical team held their breath for visible signs of life. They were shocked to witness the smaller puppy head yawn while the larger dog appeared puzzled at the new addition to its body. Further tests were made, proving both heads had the faculties of sight, hearing and smell. The puppy appeared to retain its own personality, playfully growling at the surgeons and would even lick a caressing hand!

Such stunning results prompted Demikhov to unveil his achievement to a wider audience. Keen to show off his creative genius to his peers he paraded the sensational creation before the nation's press at a meeting of the Moscow Surgical Society. Wide-eyed reporters and startled surgeons stared transfixed as one head nipped at the other before them. Heads were turned as both dogs lapped from a bowl of milk and several stomachs followed as they saw the milk dribble out of the puppy's unconnected oesophageal tube.

Unsurprisingly, Demikhov came in for some severe criticism. But he was quick to defend his actions, insisting it was not a publicity stunt but part of a long-term endeavour to discover how damaged organs can be replaced. The outlandish experiment had proved that even vulnerable brain tissue can function after being transplanted. And reporters were keen to point out this monstrous exhibit also proved Russia's superiority in the field of physiological science.

A Head Start on the Future

This was not the end of Demikhov's ground-breaking tests. It seemed the Russian scientist was not prepared to let sleeping dogs lie as over the next 15 years he created a total of 20 versions of his two-headed hounds. Nicknamed 'Surgical Sputniks' by the press, none lived for very long, succumbing to problems of tissue rejection and post-operative infection. The longest recorded survival of a two-headed dog was 29 days.

Producing these mongrel mutations was a step towards his ultimate goal: to perform a human heart transplant. Sadly for Demikhov, another surgeon beat him to it. In 1967 Dr Christiaan Barnard successfully transplanted the heart of Louis Washkansky who lived with his new ticker for 18 days. The South African doctor had visited Demikhov in 1962 to view his experiments and later credited the Russian with having inspired his work.

This Frankensteinian physiologist certainly led the way for later successes in organ transplant surgery and he continued to think outside convention – and then some. He envisioned a future where refrigerated tissue banks, containing every conceivable part of the human anatomy from corneas to kidneys, would aid transplantation. Brain-dead patients would have extra sets of limbs grafted onto them to supply amputees.

This crossing into the realm of distaste undoubtedly harmed Demikhov's career. Forever associated with his two-headed dogs he never received the proper recognition for his contribution to transplantation surgery. However, just before his death in 1998 the 82-year-old scientist was awarded the Order for Services Rendered to the Country, Third Class.

The Frozen Dead (1966). Frozen alive for 20 years! Now they return from their icy graves to seek vengeance! A mad scientist keeps the severed heads of Nazi war criminals alive until he can find appropriate bodies on which to attach them and revive the Third Reich. (Above) Dr Norberg (Dana Andrews) talks to Elsa Tenney (Kathleen Breck) whose severed head has been kept alive for the evil scientist's re-animation experiments.

HEAD TRANSPLANTS AND BRAIN SURGERY

Robert White (1926 – 2010). As East and West vied for post-war supremacy in every field of endeavour, a God-fearing neurosurgeon was made to feel the icy chill of the Cold War when called upon to indulge in state-sanctioned medical madness. Using his creative and scientific genius, Dr Robert J. White extracted mammalian brains and kept them alive using pumps and tubes, and routinely performed complete monkey head transplants before his simian science project outraged public opinion.

A Poor Catholic Boy

This highly honoured and world-renowned neurosurgeon emerged from humble beginnings. Born on 21 January 1926 Robert Joseph White grew up in a poor district within the port city of Duluth, Minnesota. Together with his three brothers, he received a strong Catholic upbringing and despite the family's destitution managed to excel at high school, graduating as valedictorian.

By the time he received his degree, his country had entered the conflict in Europe. He soon found himself serving in the South Pacific fighting a war that had killed his father. After two years of working as an Army medical technician he was discharged as a staff sergeant, aged just 19.

With his war over, Robert could now focus on further education. But with no money to speak of he was forced to rely on scholarships to fund his medical studies. He finally entered the University of Minnesota in 1949 and later transferred to Harvard Medical School where he earned his degree in 1953. Next stop was Boston, Massachusetts where he trained as a neurosurgeon at the Peter Bent Brigham Hospital. It was here he met a young nurse, Patricia Murray, whom he would later marry. Yet it was the relationship with his Government that was to have a profound effect on his career.

Cold War Competition

With the Cold War at its chilliest, the US Government watched with envy as Russian scientist, Vladimir Demikhov paraded his monstrous creations – the two-headed dogs – for all to see. America had to act quickly to challenge the current Soviet supremacy in medical science. In a bid to compete they approached Dr Robert White, assisting him in establishing a specialist laboratory at the Metropolitan General Hospital in Cleveland, Ohio. Giving him full financial backing, the authorities told the neuroscientist to outdo the Russians and put the USA back on top.

White was inspired by the work of Demikhov but he knew his dogs were little more than publicity stunts. He wanted to go further, beyond simply the stitching together of two bodies. What if he could perform a true head transplant? Surely that would place America back in the ascendancy? Knowing he lacked sufficient knowledge of brain function to complete such an audacious plan, White embarked on a journey of research and discovery.

The doctor operated on hundreds of patients with an array of cerebral injuries and diseases in his endeavours to unlock the mysteries of the brain. Yet his first medical breakthrough was spinal rather than cerebral. He developed a technique for cooling the spinal cord, slowing down the damage so surgeons could operate successfully. Next, he applied this method to the brain, pioneering new ways of chilling injured parts of the cerebrum for surgery. But this work was soon overshadowed by his more extreme experiments.

The Dog With Two Brains

After countless tests into brain activity, Dr White believed he was closer than ever to achieving the impossible. In 1962 he succeeded in making history, becoming the first doctor to successfully remove an animal's brain and keep it alive. This method known as extra-corporeal perfusion was a surgical breakthrough and would go on to be copied in hospitals and clinics around the world.

White did not stop there. He continued pushing the boundaries of cerebral surgery, attaching an isolated brain to the blood vessels on a dog's neck. In successfully keeping the second brain alive without interfering with the original brain inside the host hound, White believed he had proved this vital organ could be transplanted without fear of rejection by the body.

Yet the dog with two brains was merely a warm-up for what was to come. He began experimenting with over a hundred monkey heads, perfecting his techniques in perfusion, keeping their brains alive in specially-created solutions for up to 22 hours. Held by clamps, the macabre sight of an isolated brain maintained by tubes that flooded the organ with blood brought applause and approbation from his Government backers. However the powers that be wanted more. With the brain removed from the skull they were unable to tell whether it was truly alive. To clear up this question of consciousness, they asked Dr White to develop a technique whereby they could confirm the brain was indeed awake.

Too Much Monkey Business

The inventive doctor quickly realized in order to satisfy his state sponsors he must attempt the unthinkable: a full head transplantation. If successful this experiment would more accurately show the consciousness of the test subject. As a devout Catholic, White struggled with the moral implications of such a venture but his desire for scientific progress won out.

After several years of research and a series of complex trials, he was ready. Aided by a team of 18 bio-surgeons, technicians and nurses, Dr White prepared to make history. First, the entire head from one rhesus monkey was removed, its brain maintained using the neurosurgeon's innovative perfusion method. The next phase involved plumbing in the primate's head to the new body. The separate respiratory systems, carotid arteries and jugular veins were connected before the final stitching of the skin. The gruesome operation took six hours in total.

Then came the agonizing wait. It took nearly two hours for the pieced-together primate to wake from the anaesthetic. Suddenly, needles flickered on instrument panels and the eyelids began to move. When the eyes opened, it was clear to see the creature was far from happy.

Visually traumatized, the monkey bared its teeth at the surrounding surgical team who began yet more tests. When they probed the head with tweezers the monkey attempted to bite the offending article, it even reacted to the ringing of a bell. They quickly discovered the transplanted head possessed a full range of faculties. As well as responding to the external stimuli, the

Neurological research has come a long way since the early days of experiments on living monkeys. In 2011, a brain implant was developed that could restore memory in Alzheimer's disease. The 'magic switch', a tiny electrical impulse device, stimulates the brain and can bring back long-term memory which has been completely lost. The implant has raised hopes for millions of Alzheimer's sufferers.

primate was able to chew food and swallow although the masticated morsels travelled down an unconnected oesophagus. The results were overwhelming, the experiment had been a resounding success.

Last Frankenmonkeys

Following this astounding achievement, Dr White repeated the procedure on yet more primates. Further 'Frankenmonkeys' were created in order to perfect the technique and, while they failed to live beyond a few days, each case did prove to White that he could transplant consciousness. It was time to tell the world, revealing his results at an international neurological congress in Tokyo in October 1973. Here he declared that this advanced surgery on a monkey could actually be done on a human, stating it would actually be an easier procedure due to the larger head.

Dr White was quick to point out there were practical applications to the shocking procedure. Even without the elusive knowledge of nerve repair, ensuring any head given a new body remained paralysed from the neck down, it could still benefit those whose bodies were diseased or dying. Despite such value to medical science, his tests were condemned by the scientific community as grotesque and he was publicly denounced by animal rights activists for being nothing more than a morally-ignorant butcher.

Following such intense criticism, the US Government were pressured into pulling their funding. The Russians offered to provide him with laboratory space to continue his experiments but were unwilling to stump up the US$10 million needed. This brought an end to his brain research. There would be no more monkeying around for Dr White.

Accolades and Honours

Unfairly portrayed as a mad scientist with delusions of grandeur by the press and his detractors, Dr White always showed concern regarding the ethical implications of his work. He stood by his experiments, insisting they were not cruel but critical to advances in brain surgery. He became actively involved in topics such as IVF treatment, stem cell research and brain death.

His stance on such subjects brought him to the attention of Pope John Paul II to whom he became an adviser. In the 1980s the pontiff asked White to outline a proposal for a Vatican Commission to establish moral guidelines for doctors working in contentious fields as transplantation and genetic engineering. For a Roman Catholic who attended church every day and prayed before surgery this close association with the Holy See made him extremely proud.

Yet this was by no means the only accolade he received in his lifetime. Honorary doctorates and medical awards came thick and fast to the man compared by many to Dr Frankenstein. In 1997 he was given the Humanitarian Award from the American Association of Neurological Surgeons, acknowledging his contribution to the advancements in brain surgery. A year later he retired, moving to Geneva County, Ohio on the banks of Lake Eerie. On 16 September 2010 the father of 10 and grandfather to 20 died after suffering from diabetes and prostate cancer.

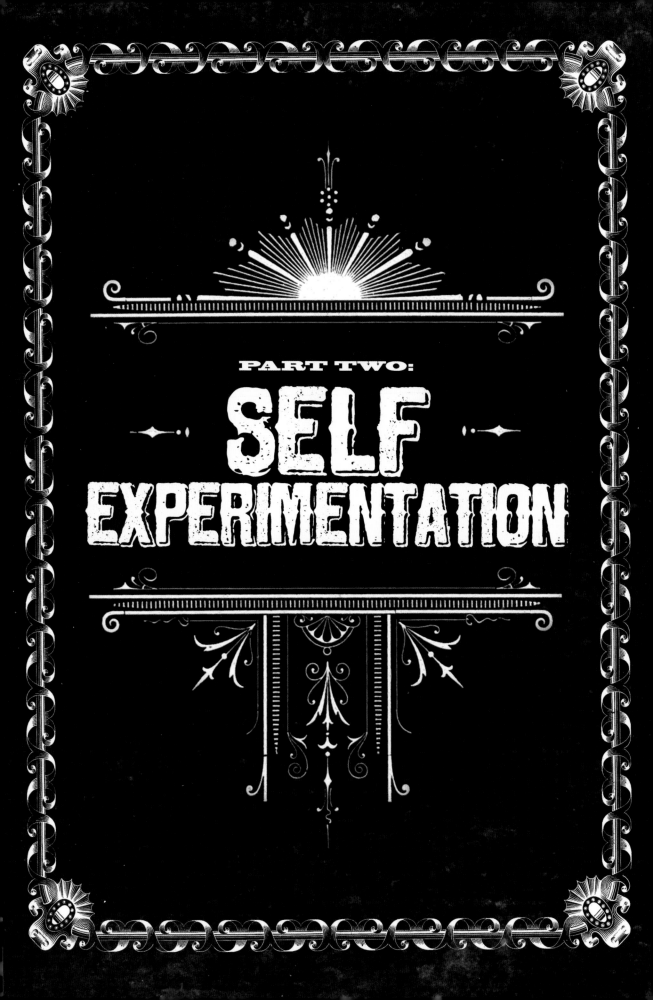

PART TWO:

SELF EXPERIMENTATION

CURING THE YELLOW SCOURGE

Stubbins Ffirth (1784 – 1820), is most widely remembered as the doctor who drank infected vomit and poured it into his eyeballs while trying to prove that Yellow Fever was not contagious. To discover the cause, Ffirth also tried smearing his body with infected blood, saliva and urine. True, Ffirth's methods may have been a little unorthodox, but they came at a time when yellow fever was reaching epidemic proportions, and no-one, not even the best physicians in the land, had a clue how to deal with it.

In 1793, the port city of Philadelphia was enjoying something of an economic boom. The rising demand for tobacco and sugarcane was attracting new wealth into the town, and the ongoing Haitian Revolution had forced many French refugees to claim it as their home. However, these halcyon days were not destined to last for long.

Unfortunately the native population failed to realize that many of these Haitian immigrants were suffering from a disease known as the Scourge of the Americas. It was yellow fever, the symptoms of which include a high temperature, nausea and intense pain. As the disease progresses it causes liver damage with jaundice, giving the patient's skin a ghostly yellow pallor and giving rise to the name of the disease. Death is the final outcome. The 1793 outbreak was destined to become one of the worst epidemics in US history, causing the deaths of 10 per cent of the Philadelphian population within the first month alone. On average it killed one in five sufferers, not to mention threatening the social and economic development of a whole section of North America.

Philadelphia's physicians had literally no idea how to combat Yellow Fever. They did not know what caused it, how it spread or, indeed, how to treat it. Most doctors of this time believed in the healing power of nature. They thought that the body would eventually rid itself of any toxin or illness, and that a doctor's job was simply to aid this process as much as possible. But this approach simply wasn't working and the residents of Philadelphia found themselves in a state of panic. The almshouses were full to bursting and the streets were littered with the sick and the dying. The populace desperately needed an alternative treatment. Luckily for them one or two brave souls were willing to have a go.

The Prince of Bleeders

The Founding Father and physician Benjamin Rush saw that the process of simply leaving patients to die was not working terribly well, and so he set about trying to find a cure. He tried all sorts of remedies, from brandy and tree bark infusions to wrapping the patient up in a blanket soaked in warm vinegar but, perhaps unsurprisingly, none of these worked. So, he returned to his books and found a letter from a Dr John Mitchell, in which Mitchell had described his own method of treating patients during an earlier outbreak of yellow fever. He recommended purging the stomach and intestines by whatever means necessary, urging doctors to dismiss 'any ill-timed scrupulousness about the weakness of the body' and get on with it. As a result of this discovery Rush revived a purging treatment called 'Ten and Ten', which involved administering 10 grams of mercury and 10 grams of the cathartic drug jalap, only his treatment had one major difference. In order to accelerate the process he upped the dose to 10 and 15 grams, and administered it three times a day, basically poisoning the patient to get rid of their stomach contents. He was, in effect, making his patients far worse in order to eventually cure them. Rush was also, like most of his contemporaries, a passionate believer in bloodletting and routinely removed so much blood from his patients that they passed-out, earning him the nickname 'Prince of Bleeders'.

Despite the brutal nature of his treatments, Rush said he saw a marked improvement in eight out of 10 patients, and even claimed that many recovered after only one session. So certain was he that, when he finally contracted the fever himself, he had a servant administer his own treatment and, sure enough, he recovered within five days. Still, his methods attracted a lot of criticism and many of his peers openly accused Rush of poisoning those under his care, which of course, he was. Rush's research also failed to answer the most crucial of questions; how to stop people catching yellow fever in the first place?

Drinking the Black Vomit

Of all the deeply unpleasant symptoms of yellow fever, the vomit seemed to strike particular fear into the hearts of the populace. This is probably down to its alarming appearance. In cases of yellow fever, the liver becomes damaged and affects the blood's ability to clot, and so blood fills the stomach, which has to be evicted somehow. The result is copious amounts of thick, pitch-black vomit. In 1800, an American physician named Dr Isaac Cathrall decided to perform some experiments involving the substance in question, an undertaking he considered extremely hazardous. He repeatedly put black vomit from several patients to his lips and tasted it. It must have been a horrible experience, but crucially, he did not develop yellow fever.

Ffirth's Feverish Theories

A few years later Stubbins Ffirth was training to become a doctor at the University of Pennsylvania when he decided to focus his research on the disease that had so decimated the local population. He recognized that there were far fewer instances of yellow fever in the winter than during the hot summer months. He theorized that this was because the disease was caused by strenuous weather conditions. By careful research and observation he noticed that the doctors and nurses who treated sufferers did not

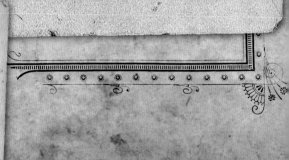

A Yellow Fever patient (1910) inside a portable isolation cage at Ancon Hospital during the construction of the Panama Canal. Ancon Hospital played a crucial role in controlling yellow fever, malaria, and other tropical diseases during the period of the excavation of the canal (1904–1914). William C. Gorgas, Chief Sanitary Officer founded the hospital and his achievements in the field of tropical medicine were recognized in 1928 when the establishment was renamed Gorgas Hospital. After more than a century of clinical tropical research activities in Panama, Gorgas Hospital closed its doors in 1997.

routinely fall victim themselves. Neither did the family members, the hospital attendants or even the gravediggers who were supposedly exposed to it. Inspired by this, and Cathrall's earlier research, he set out to prove once and for all that yellow fever was not contagious. It's not entirely clear whether he chose to experiment on himself as an expression of self belief, or because no one else was reckless (or stupid) enough to volunteer themselves, but experiment he certainly did.

Ffirth was determined to prove that yellow fever was not contagious and his endeavours were based on good common sense. He thought if people understood that they were unlikely to contract yellow fever from their sick relatives, they would be less likely to abandon them. Patients who were properly cared for were much more likely to survive. Many yellow fever fatalities died due to aspirational pneumonia caused by inhaling their own vomit. This was easily preventable with proper nursing care. If Ffirth could prove that yellow fever was not contagious, it would cut the costs of the enormous quarantines which were set up to contain the virus, but succeeded only in damaging the economies of the towns and cities that were already suffering at the hands of the fever.

Blood, Sweat and Body Fluids

Ffirth began by experimenting on animals. He fed cats and dogs on vomit for one week. When they survived, he cut open the skin of a dog, inserted the black vomit and closed the wound back up again to prevent it from escaping. Still, the animal lived. Spurred on,

Ffirth looked to himself as a further source of experimentation. In 1802, Ffirth decided to expose himself to the disease in as many ways as possible, beginning by bringing himself into direct contact with the bodily fluids of sufferers. He made incisions in his arms and smeared the wounds with vomit. He dribbled the stuff onto his eyeballs. He even fried some up on a skillet and inhaled the fumes. When he failed to contract the disease, he turned to drinking it undiluted – still nothing. At this point most people would have shut the experiment down and declared themselves victorious, but not Ffirth. He wanted to prove that other bodily fluids were as harmless as the vomit was, so he smeared his body with blood, sweat, saliva and urine. He still managed not to contract the disease. It later turned out that Ffirth's samples came from patients in the later stages of yellow fever and they were no longer contagious.

The Fly in the Ointment

Although he never succumbed to yellow fever, ultimately Ffirth's hypothesis was flawed. It is true that yellow fever is less common during the winter months, but this is because the mosquitoes that carry the disease die off in the winter. Yellow fever is actually extremely contagious, but can only be caught through blood to blood contact. A further 60 years passed before a Cuban scientist named Carlos Finlay finally discovered the link between yellow fever and mosquitoes. Although it can now be prevented, there is still no effective treatment for yellow fever once the illness sets in.

COCAINE RUNNING ROUND MY BRAIN

William Halsted (1852 – 1922). The father of American surgery, William Halsted was an inspirational teacher, natural innovator and a competitive and daring self experimenter. Ironically, Halsted's unquenchable thirst for knowledge threatened to end his sparkling medical career when he and his research team set out to discover if cocaine could be used as a local anaesthetic. Within weeks the good doctor was transformed into a hapless drug addict. And he wasn't the only one.

William Halsted had it all. Born in 1852 in New York City into a wealthy and successful family, he was tutored at home until the age of 10, and then sent off to boarding school in Massachusetts, where his earliest interests were of a sporting nature. It wasn't until his last year at Yale that this captain of the college football team began to consider medicine as a career. For most people, becoming a doctor is something one has to work long and hard for, and yet it all seemed to come so easily to him. From Yale he went on to Columbia University College for Physicians and Surgeons and from there to Europe, where he was lucky enough to study under some of the world's most prominent surgeons. It must have seemed as if success was just ripe for the plucking, all he had to do was reach out and grab it. Unfortunately, life had other ideas.

Cocaine

Cocaine today is seen as a sleazy, illicit street drug. A guilty pleasure for the rich and famous. Things were different in the 19th century. Cocaine was big business, and very much part of the mainstream. Vin Mariani, a Bordeaux wine treated with coca leaves, was the tipple of choice for many of society's elite. Thomas Edison and Queen Victoria were both big fans, as was Pope Leo XIII, who liked the drink so much he awarded it a Vatican gold medal and even appeared on a poster advertising it.

While Vin Mariani flowed from the decanters of the rich and famous, drug companies were hailing cocaine as a wonder drug that could be used to treat all manner of ailments, including fatigue, migraines and post-operative pain. But, for all their enthusiasm, the medical profession didn't

An 1885 poster from Berlin, Germany, advocating the
new cool craze for taking cocaine.

know much about it. They raved about the benefits but utterly failed to understand the rampant addiction it caused.

Freud's Coke Compulsion

Long before Sigmund Freud discovered psychoanalysis, he discovered cocaine. In fact, his critics sometimes cite his cocaine use as the reason behind some of his theories. But Freud wasn't just a user, he was an evangelist. Not content with doing the odd bit of coke himself, he recommended it to friends and loved ones. He even sent some to his fiancé Martha Bernays, along with a letter that read:

I will kiss you quite red and feed you till you are plump. And if you are froward* you shall see who is the stronger, a little girl who doesn't eat enough or a big strong man with cocaine in his body. In my last serious depression I took cocaine again and a small dose lifted me to the heights in a wonderful fashion. I am just now collecting the literature for a song of praise to this magical substance.

The 'literature' he referred to is his paper *On Coca*, which he published in 1884. Between 1883 and 1887 Freud wrote a number of articles recommending the medical use of cocaine and promoting it as an antidepressant amongst other things. He believed that he could use it to help cure his friend Dr Ernst von Fleischl Marxow of a morphine addiction he'd developed following a painful infection. The results were nothing short of disastrous. He succeeded only in causing Marxow to develop an acute case of cocaine psychosis. Ultimately he turned back to morphine, and died a few years later in extreme pain.

Freud used cocaine compulsively for 12 years, continuing to use it even after stories of addiction and overdose began surfacing and he stopped recommending it to others. By then, though, the damage had already been done. The drug had succeeded in destroying some of his closest relationships and tarnishing his reputation. He gave it up finally in 1896.

Coca Koller Jumps to Conclusions

Freud's promotion of cocaine helped to convince his colleague, the 26-year-old eye surgeon, Karl Koller, that it could be used to anaesthetize the surface of the eye. Before Koller's experiments with cocaine, eye surgery, which often requires the patient to be awake, was dangerous, painful and extremely difficult. It had to be performed without adequate anaesthesia; the patient was required simply to lie still and resist the urge to flinch as the surgeon's scalpel came towards him. An impossible task, as the eye is sensitive to even the most minor stimulus. To begin with Koller and his team experimented on a frog. One of the researchers who took part made these notes:

A few grains of the substance were thereupon dissolved in a small quantity of distilled water, a large lively frog was selected from the aquarium and held immobile in the cloth, and now a drop of the solution was trickled into one of the protruding eyes. At intervals of a few seconds the reflex of the cornea was tested by touching the eye with a needle ... After about a minute came the great historic moment, I do not hesitate to designate it as such. The frog permitted his cornea to be touched and even injured without a trace of reflex action or

* 'froward' is the old-fashioned word Freud used, meaning 'difficult to deal with'

attempt to protect himself – whereas the other eye responded with the usual reflex action to the slightest touch.

Bolstered by this result, Koller and his team began sticking pins in each others' eyes to observe the effect first hand.

An Addiction that Changed the World

Reading about Koller's experiments William Halsted wondered how this numbing effect could be used in general surgery. Later that year, he and his associates in the Outpatient Department of Roosevelt Hospital and 30 medical students embarked on a study to find out exactly how useful cocaine could be. The answer, incidentally, turned out to be 'very', if you can somehow bypass the dramatic mood swings, heart palpitations, anxiety, panic attacks, dizziness, nausea, not to mention the life endangering and altering addiction that comes with it.

Tragically for Halsted and his team, cocaine addiction rapidly took hold, turning what began as a serious scientific study into little more than a farce. The experiment's participants began using cocaine freely and indiscriminately; socially as well as medically, injecting and snorting the stuff both inside and outside of the laboratory. The people around him had turned from a highly respected circle of doctors into an unpredictably rabble of drug addicts.

Halsted would battle cocaine addiction for the rest of his life, sometimes substituting coke for morphine, or alcohol, but always returning to it sooner or later. His work suffered, his once steady surgeon's hands became shaky and untrustworthy, and so he withdrew from the operating table. The popular and charismatic doctor had become a shadow of his former self, a social recluse who often took long vacations alone, his precise whereabouts unknown to even those closest to him. When he was on home turf, whole weeks might go by without his appearing at the hospital. He preferred to work at home rather than deal with interruptions caused by actual patients.

Halsted versus Reality

Despite suffering from a crippling drug addiction, it must be said that Halsted did manage to accomplish some incredible things. He was responsible for founding the first surgical residency training programme in the United States. He introduced the radical mastectomy for breast cancer sufferers, which has no doubt saved millions of lives. He even introduced the latex surgical glove to operating theatres – before he began wearing them, most surgeons worked with their bare hands. His contribution to medical science cannot be underestimated, but one can't help but wonder what else he might have achieved, how many more lives might have been saved, if it had not been for the debilitating drug habit he acquired in the course of his research.

THE CARDIAC KID

Werner Forssmann (1904 – 1979). How much courage does it take to insert a catheter into your own heart? Imagine you are the first person ever to do it, your colleagues think it will kill you and your boss forbids it. In 1929, Werner Forssmann did just that, during an experiment to develop a technique of catheterization of the heart. What made him gamble his own life to prove a point? Did it come down to a mature sense of social responsibility, or was it just sheer madness?

In order to become a successful and ground-breaking scientist one must be brilliant enough to have the idea in the first place, *and* brave (or mad) enough to test it out. Most people are capable of dreaming up fantastic things, but relatively few possess the passion and self-belief to go one step further and actually try it. Of those few, still fewer are willing to use their own bodies as a laboratory instrument. Werner Forssmann was one of those rare individuals. But what exactly did his self-experimentation bring him, besides a shared Nobel Prize?

Werner Forssmann was born in Berlin in 1904, the son of a German solicitor. He studied medicine at the University of Berlin, and received his MD degree in 1929. After graduation he went to work at the August Victoria surgical clinic on the outskirts of the city and it was here that he formulated his theory for cardiac catheterization.

Forssmann thought that if he could insert a thin, flexible tube (a catheter) into the heart via the groin or the arm, he might be able to deliver drugs to a patient much quicker in emergency situations. Up until then, the same effect could only be achieved with a blind needle injection through the chest wall – an incredibly dangerous procedure.

If he could catheterize the heart, he would also be able to measure the volume, rate of blood flow, pressure and oxygen content within it, which would help improve doctors' overall understanding of the organ. In those days, most believed that a catheter would become entangled in the chambers of the heart and cause it to stop beating, resulting in instant death.

But 25-year-old men can be incredibly foolhardy, and Forssmann was no exception. His bosses' refusal to sanction his experiments acted like a red rag to a bull and he resolved to prove them wrong, even if it cost him his career, and his life.

Cadavers on the table

First of all Forssmann set about practising his technique on the dead, proving to his satisfaction that he could safely pass a catheter from an incision in the crook of the arm, up a vein and into the heart. That's all well and good, but, because a dead person's heart has a tendency not to beat, he still really had no idea what the effects on a living person might be. He needed to go one step further.

Next, Forssmann managed to convince a fellow resident doctor to help him with the initial stages of his self-experiment. Secretly one night his colleague inserted a large needle into a vein in Forssmann's arm. After which Forssmann continued himself, inserting a catheter and advancing it towards his ticker.

Unfortunately for Forssmann, his colleague got cold feet. At the last moment, fearing he was about to become involved in a suicide, his accomplice refused to continue. Forssmann was forced to quit his experiment prematurely.

The Main Event

Forssmann really was a man on a mission, so it will come as no surprise whatsoever that this setback did nothing to diminish his determination. A week later he managed to convince a nurse named Gerda Ditzen to assist him, and had another stab at it.

It's not entirely clear why she agreed to help him out. In his autobiography, Forssmann relays a rather colourful account of events in which he charms the beautiful nurse into thinking he is anaesthetizing her arm, when he is really anaesthetizing his own. By the time Gerda had realized what she'd got herself into, it was too late. He tied her to a chair, and she was left no choice but to watch while Forssmann heroically performed his groundbreaking experiment.

Many historians doubt that the real events were quite as melodramatic. But it is true that a nurse agreed to assist him. Perhaps she was just curious, or perhaps she thought someone had better be on hand in case the silly fool really did meet his maker.

After painstakingly pushing the catheter an astonishing 63 cm up his cephalic vein, through the bicep and past his shoulder and subclavian vein, Forssmann guessed it may have reached his heart, and walked downstairs to the X-ray room with the rest of the tube dangling from his arm. He asked the radiologist to take an X-ray as proof that he had been right all along. The end of the catheter had entered the right atrium of his heart. In response, the

Dr Werner Forssmann, photographed in 1956, the year he was jointly awarded the Nobel Medicine Prize with Professor Andre F. Cournand and Professor Dickinson W. Richards of the Columbia College of Physicians and Surgeons. Three of the world's outstanding heart specialists, they were jointly awarded the prize for their discoveries in heart catheterization.

hospital recognized his discovery, but removed him from his post.

Later that year, he published the results of his experiments, suggesting that the technique could be used to inject dye into the heart to obtain better heart X-rays, to help diagnose heart defects and measure blood pressure inside the heart.

However, Forssmann was unpopular with his colleagues and contemporaries. They wrote off his research as little more than attention seeking by a young doctor with much to learn about hospital politics. Bitterly disappointed and frustrated, Forssmann rejected cardiology altogether and became a urologist.

It was not until after World War II, a period during which he was an active member of the Nazi Physicians' League and a high-ranking medical officer in the German army, that Forssmann's research into catheterization of the heart attracted any attention at all.

What did it mean for humanity?

In the early 1940s, two scientists at Columbia University named Dr Andre F. Cournard and Dickinson W. Richards, stumbled across the paper in which Forssmann described his self-catheterization experiments. For them, it lit the way towards the development of safe and practical methods for catheterizing the heart and the lungs. As a result, both Forssmann, and Cournard and Richards shared the 1956 Nobel Prize for Medicine.

Apparently, when he was told he had won a Nobel Prize, Forssmann replied 'for what?' The experiment had evidently been forgotten even by the man who risked his life to perform it. In the same year, Forssmann was appointed Honorary Professor of Surgery and Urology at the Johannes Gutenberg University, Mainz. His misdemeanours seemingly forgotten, a string of other awards and accomplishments followed.

Today, cardiac catheterizations are commonplace throughout the world. If it were not for the mad youthful self-experiments of Werner Forssmann, everyday life-saving surgical techniques such as coronary angioplasty, may never have been possible.

The historic X-ray taken to show the first cardiac catheterization experiment that Werner Forssmann performed on himself in 1929.

THE KALEIDOSCOPIC COLOURS OF LSD

Albert Hofmann (1906 – 2008), invented LSD and took the drug many times but he never meant the rest of the world to 'tune in and turn on'. He saw it as a gift from the universe to the human race, an evolutionary tool that would help elevate the world to the next level of consciousness. Others saw it as a way to counterbalance the negative karma generated by the development of the nuclear bomb. Whatever the universe had in mind, on 19 April 1943, Hofmann went for an extremely interesting bicycle ride, and humanity has never looked back.

Albert Hofmann was born in Switzerland on January 11 1906. His father, Adolf Hofmann, was a factory toolmaker, and the family's low income meant that Albert's godfather paid for his education. At 20 years old, Hofmann took up a place at the University of Zurich, where he studied for a chemistry degree. His interest in chemistry stemmed from a fundamental philosophical question: Is the material world a manifestation of the spiritual world? Hofmann hoped to find sound answers from the solid laws of chemistry, and to apply these answers to the problems of life. These answers, as far as Hofmann was concerned, would be found within the plants and animals that populated the planet. In his own opinion, and those of his most devoted fans, his invention of LSD, a drug that operates at the deepest level of human consciousness, even in the most miniscule amounts, proved him right.

Discovering the Mind Bender

With a doctorate under his belt, Hofmann went to work for the pharmaceutical- chemical department at Sandoz Laboratories, which is now a subsidiary of Novartis. He began studying the medicinal plant squill and the notoriously psychoactive fungus, ergot, as part of an attempt to synthesize their active components for medicinal use.

One of these constituents, the 25th lysergic acid derivative synthesized by Hofmann, later became known throughout the world as LSD. It passed completely under the radar to begin with because it had no obvious effects on the animals he tested it on. As a result, Hofmann set it aside.

It was not until 1943 that Hofmann realized he may have missed something and returned to re-examine it. He synthesized LSD a second time, and even then the discovery of its mind-bending properties happened more by luck than by judgement. He must have accidentally absorbed a tiny quantity of LSD through his fingertips, because soon afterwards he was forced to leave the laboratory and return home to regroup. Hofmann didn't realize it at the time, but his notes describe the first LSD trip in human history:

> *Last Friday … I was forced to interrupt my work in the laboratory in the middle of the afternoon and proceed home, being affected by a remarkable restlessness, combined with a slight dizziness. At home I lay down and sank into a not unpleasant intoxicated-like condition, characterized by an extremely stimulated imagination. In a dreamlike state, with eyes closed … I perceived an uninterrupted stream*

Swiss scientist Albert Hofmann was the first person to synthesize, ingest and learn the psychedelic effects of LSD. Photographed in 1988.

of fantastic pictures, extraordinary shapes with intense, kaleidoscopic play of colors. After some two hours this condition faded away.'

Hallucinogenic Bicycle Trip

Three days later, on 19 April 1943, Albert Hofmann made the momentous decision to intentionally take 250 micrograms of LSD. He predicted this to be the drug's threshold dose (the lowest effective dose) but, in fact, 20 micrograms is the threshold dose for LSD. 250 micrograms was actually an enormous dose, and so, less than an hour later, he began to experience wide-ranging shifts in perspective. Feeling peculiar and worrying that he might have made himself ill, Hofmann rode 6 km home on his bicycle.

On the journey, his condition deteriorated rapidly. He struggled with feelings of anxiety, worried that he was going insane and wrestled with paranoia, at times even believing that his neighbour was an insidious, evil witch with a coloured mask. When a doctor arrived to examine him, however, he could find no physical abnormalities, save for a pair of massively enlarged pupils, of course. Eventually Hofmann's terror began to recede, making way for feelings of enjoyment and elation:

Little by little I could begin to enjoy the unprecedented colors and plays of shapes that persisted behind my closed eyes. Kaleidoscopic, fantastic images surged in on me, alternating, variegated, opening and then closing themselves in circles and spirals, exploding in colored fountains, rearranging and hybridizing themselves in constant flux...

From this day Hofmann was convinced that LSD would have a fundamental impact on the world, and particularly on the human race. Over the course of his career he came to regard it as a tool given to human kind in order to help us become what we should be – a catalyst for positive change – a kind of evolutionary WD40. To Hofmann, LSD proved there is an inseparable interaction between the material and the spiritual world.

He would become an advocate for its use, both in the arenas of personal development and formal psychoanalysis, but first he needed to convince his boss, Professor Stoll, the head of the Sandoz pharmaceutical department, as well as the head of the pharmacological department. They didn't believe him. They said he must have made a mistake with the dosage, that it was impossible that such a small quantity could have that kind of effect. But when others in the pharmacological department at Sandoz repeated his experiments, they found that he was right about the incredible potency of LSD. Even lab staff who took just 50 micrograms reported very strong hallucinogenic experiences.

Like the head honchos at Sandoz, the rest of the world also needed convincing, so it was made freely available to qualified clinical investigators. The properties label on the bottles read:

Caution – Causes hallucinations, depersonalization, reliving of repressed memories and mild neuro-vegetative symptoms.

Medicine for the Soul

In the early 1950s there were numerous studies into the possible applications of LSD. It was soon being used by psychiatrists to treat all forms of mental illness including neurosis, psychosis and depression. It was even used to treat alcoholism – in fact, at least one study found it to be more effective in the treatment of alcoholism than anything else before or since.

Newspapers and magazines began to fill column inches with positive stories about LSD experiments, miraculous effects, mystical rebirths and self-transformations. Even movie star Cary Grant experimented with LSD-assisted psychotherapy, claiming afterwards that it made him feel as if he had been reborn. Psychotherapists began to use the drug themselves and to distribute it amongst their friends.

Eventually, LSD became the drug of choice for anyone who wanted to fully realize their creative potential, whether from the comfort of a therapist's couch, in their own home, at a party or music concert. A counterculture had been born, but not everyone was happy about it, Hofmann least of all. He had never imagined that LSD would be used for recreational purposes. To him, it must have been akin to ripping a page from a copy of the Bible and using it to light a marijuana cigarette. The universe gives you the greatest gift ever bestowed upon the human race, and what do you do? Party and write rock songs! He accused rank amateurs of hijacking the drug without fully understanding its positive or negative effects.

Rampant and irresponsible use of the drug inevitably led to bad trips and mental breakdowns. Hofmann could see that it was being widely misused and this made him angry, but there was an additional problem. He still believed that LSD, if taken in the right way, could help answer the most fundamental questions life had to offer. But, it provided the user with a new concept of life, a way of looking at existence that was opposed to the officially accepted view, and this made it dangerous. He worried that governments would grow scared of its potential to wake people up to another way of being, and begin to legislate against it. In this, at least, he was right.

Aftermath of the Outlaw

In 1966, despite the fact that it was neither toxic nor addictive, the drug was outlawed across the world. It remained available for strictly monitored psychiatric use, but the people who wanted to use it became so wrapped up in red tape that it eventually put them off. Slowly LSD became relegated to the streets, where, because of its often troubling side effects and association with mental illness, it all but faded into obscurity.

Albert Hofmann died of a heart attack at the age of 102, and although he experimented with a number of other psychedelic substances, he remained an advocate of LSD right up to the end of his life. All he really wanted was for its medicinal properties to be fully recognized:

As long as people fail to truly understand psychedelics and continue to use them as pleasure drugs, and fail to appreciate the very deep psychic experience they may induce, their medical use will be held back. While the world has still to take full advantage of its gifts, the doors of perception will remain tightly shut.

DOLPHINS & DRUGS

John Lilly (1915 – 2001). A relentless adventurer, John Lilly journeyed repeatedly to the very edge of knowledge in order to explore the nature of reality. A committed LSD experimenter, whose friends included Timothy Leary, Allen Ginsberg and Aldous Huxley, John Lilly had more to offer the world than drug-induced ramblings of a rock and roll drop out. His work encompassed many of the over-riding themes of the 20th century, from extra-terrestrial life forms to inter-species communication.

John Lilly was born to Richard and Rachel Lilly on the 6 January 1915 at St Paul, Minnesota. His early life was a very conventional one, but, by the age of 13, a passion for science had taken hold and he was conducting his own experiments in the family basement. He studied Physics and Biology at the California Institute of Technology, graduating in 1938 before receiving his medical degree from the University of Pennsylvania in 1942, specializing in biophysics and psychoanalysis.

Once World War II was in full swing Lilly got a job at the Johnson Foundation for Medical Physics, researching the effects of high altitude flying. He invented instruments for measuring gas pressure, and after the war, he continued his training in psychoanalysis.

No Limits in the Province of the Mind

In 1953, Lilly moved to study neuropsychology with the US Public Health Service Commissioned Officers Corps. He was charged with studying the effects of sensory deprivation, and, in a bid to isolate the brain from external stimulation, he set about designing the Lilly Tank – the world's first sensory deprivation tank.

The idea was to discover whether or not the brain continues to function without outside stimulus. Lilly hypothesized that without the use of the senses the brain would enter a kind of sleep state. He used the tank to explore the origin of consciousness and its relationship with the brain. He regularly spent long periods submerged in the tank, and noted the kinds of experiences he had in there, some of which saw him transcend his physical form entirely. He began to visualize himself conferring with two spiritual guides.

Scientific visionary John Lilly, in 1977.

It was during one of these sessions that Lilly realized the US government could no longer support the kind of research he really wanted to do. They had started to impose controls on the isolation tank work and on the brain research under his remit. He would have to change direction.

Extra-terrestrial Counterparts

During the late 1950s Lilly studied large-brained mammals, in particular, dolphins. He believed that dolphins and whales are highly developed creatures that are capable of thinking and communicating in ways humans cannot, describing them as extra-terrestrial counterparts that hail from the same planet.

Cetaceans have brains that are at least as big as a human brain, and they have had them 50 million years longer. Lilly argued that dolphins and whales deserve to be treated with the highest possible respect.

He became a lifelong campaigner for the development of communications between humans and dolphins and established a facility for developing relations between the two species. He even published papers arguing that dolphins could mimic human speech patterns.

Sadly, when his experiments were repeated, scientists were unable to replicate his results. Ignoring the science, Lilly continued to talk to the dolphins, calling the language 'dolphinese'.

Space Cadet and Psychonaut

Interested in the search for extra-terrestrial intelligence, Lilly helped gather a group of scientists at the Green Bank Observatory in 1961. They devised techniques in radio astronomy to detect alien life forms outside the solar system. Adopting the soubriquet The Order of the Dolphin, after Lilly's communication experiments, they formulated the Drake equation to estimate the number of extra-terrestrial civilizations in the galaxy.

Lilly found LSD and ketamine in the early 1960s and embarked upon a series of drug-fuelled experiments in which he ingested psychedelic drugs either with dolphins or in an isolation tank, sometimes both. He recorded many of his experiences and observations, although he said his first LSD experience in a sensory deprivation tank was way too intense to speak about. In the end, Lilly's research and writings may have had more in common with the beat poet Allen Ginsberg than with any of his scientific contemporaries.

In the 1980s, he was back with his beloved dolphins, teaching them a computer-synthesized language and designing an underwater floating living room where humans and dolphins could get together for a chat.

Lilly's vision of a time when all killing of whales and dolphins would cease, inspired him to move to the idyllic island of Maui in Hawaii where he lived surrounded by his aquatic mammalian friends until his death in 2001.

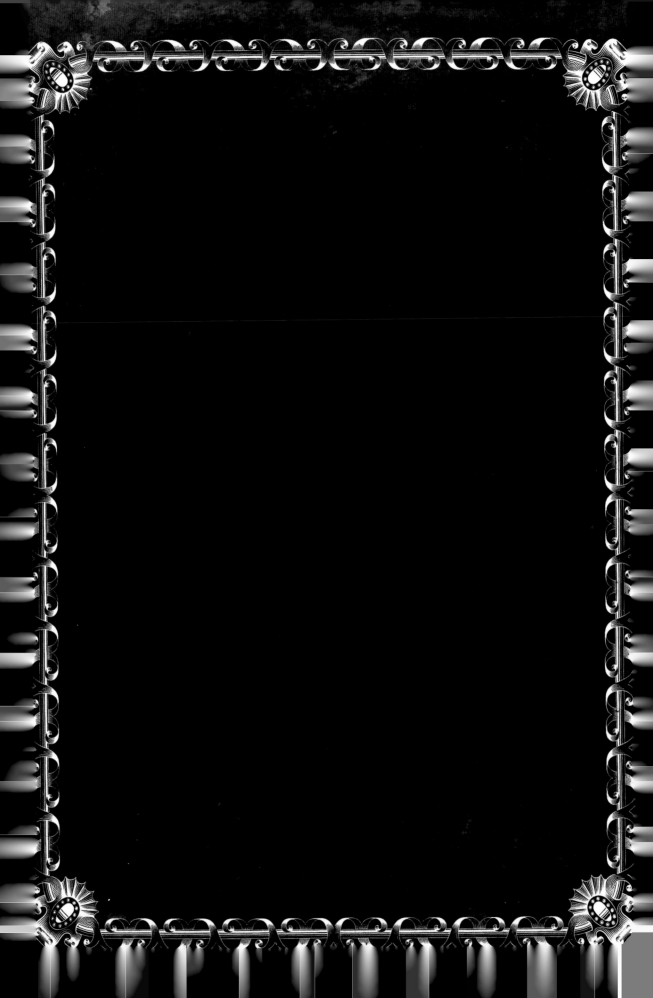

PART THREE:

MIND CONTROL

MESMERISM AND THE HYPNOTIST

Franz Mesmer (1734 – 1815). Was Franz Mesmer the father of alternative medicine, or a charlatan? Almost 200 years since his death the jury is still out. As is so often the case with people who spend their whole careers flying in the face of conventional thought, his story is a dramatic one, full of incredible highs and devastating lows.

In the 18th century, the body, its functions and frailties were still largely a mystery. Most doctors continued to base their diagnosis on the ancient and widespread belief in the four humours, which had been around since the days of antiquity and had no basis in reality whatsoever. The practice of bleeding patients with leeches was still commonplace, and medicinal cures could contain anything and everything – even bits of dead humans.

This was the world Franz Mesmer entered when he graduated with a doctorate from the University of Vienna in 1766; a world before anaesthetic or antiseptic, in which lay-healers were often more successful at curing patients than educated men. His theories concerning animal magnetism and the power of the planets in relation to the human body may seem insane now but, in a historical context, they were no more crazy than the practices of most conventional doctors of his day.

One of Nature's Healers

Franz Anton Mesmer was born in Swabia, Germany in May 1734, the son of a master forester, Anton Mesmer, and his wife Maria. Mesmer's parents were 'honest, pious and without means', and encouraged Franz to enter the church, so Franz studied law and theology at the Jesuit universities of Dillingen and Ingolstadt before moving to Vienna to study medicine at the university there. Mesmer was heavily influenced by Isaac Newton's theory of the tides.

He believed that the human body was controlled by an animal magnetism, the same gravitational pull that the moon, sun and earth exert on the world's oceans – beliefs that were not looked on favourably by his religious tutors. Disease, according to Mesmer, was caused by an imbalance in this tide, but could be cured using magnets.

Some people – nature's healers – were better conductors of animal magnetism than

Franz Mesmer, German physician and hypnotist, treating patients in a group session which he called a baquet (named after the large tub in the middle of the room), at a fashionable salon in Paris, France 1785.

others. Needless to say, Mesmer saw himself as one of these gifted souls, put on earth to channel the force to others.

The Franzl Oesterlin Problem

In 1768 Mesmer married a wealthy heiress named Anna von Posch. He settled quickly into the life of a Viennese socialite, befriending local aristocracy, royalty and celebrities including the Mozart family, opening a conventional doctor's practice and becoming a patron of the arts. It is through his friendship with the Mozart family that Mesmer's earliest animal magnetism experiments are recorded. Leopold Mozart's letters to his wife, in 1773, tell the story of a seriously ill 27-year-old woman named Franzl Oesterlin. She came to Franz Mesmer for treatment, suffering from what Mesmer described the symptoms as 'blood rushing to her head and setting up the most cruel toothaches and earaches, followed by delirium, rage, vomiting and swooning'.

Mesmer deemed the malady so serious that Franzl moved into his house to receive 24-hour care. Mesmer used magnets to disrupt the gravitational tides, which, he believed, were causing the patient's extreme symptoms. He asked her to visualize the excess fluid draining rapidly from her body, taking her illness away with it.

Her rehabilitation was not instantaneous, but she did make a full recovery. Leopold Mozart reported that 'Franzl had been dangerously ill, and blisters had to be applied to her arms and feet ... they had expected her to die... .'

Mesmer didn't know it at the time, but he had actually eased Oesterlin's symptoms not with the power of the planets, but with the power of the mind.

PHYSIQUE ET CHIMIE POPULAIRES

Le Baquet de Mesmur.

Celebrity Beckoning

Following Franzl Oesterlin's miraculous recovery, Mesmer went on to successfully treat a wide variety of patients suffering from psychosomatic blindness, paralysis, convulsions and other 'hysterical' conditions. He garnered a certain amount of celebrity, giving dramatic demonstrations in the palaces and grand houses of Europe but the medical community rejected his theories, and continued to see him as a showman rather than as a healer.

Mesmer's reputation took a further pounding when he attempted to treat an 18-year-old concert pianist named Maria-Theresa Paradis for psychosomatic blindness. Treatment was progressing well when Maria's parents, who stood to lose their royal pension if their daughter was cured, forcibly removed her from Mesmer's care, whereby her symptoms (unsurprisingly) returned.

Mesmer's doubters seized upon this unfortunate episode, regarding it as proof that he was a fraud. He subsequently left Vienna and moved to Paris in 1777. He rented an apartment in an affluent, fashionable part of the city, establishing a medical practice. He continually failed to find any other scientists or doctors who would take his theories seriously, but patients were never in short supply.

His client list included none other than Marie Antoinette herself. Even the wealthiest patients were forced to book appointments weeks in advance. In fact, his surgeries proved so popular that he was forced to devise a way to treat people without the need for so much one-to-one care.

The Master of Theatre

Mesmer couldn't resist a bit of melodrama and eschewed the idea of a quiet, private little consulting room in favour of something a lot more theatrical. He treated people individually or in groups, séance style. Atmospheric music was played on a clavichord while his patients filed into the lavishly furnished room to await the Master's performance.

Seating themselves around the baquet, a large wooden tub filled with a mixture of iron filings, powdered glass and magnetic water, his patients were asked to grasp magnetic rods to facilitate the transfer of the animal magnetism. The lights were dimmed and incense was used to heighten the drama. Mesmer then made a grand entrance dressed in a flowing silk robe and carrying a long iron wand.

Circling the room, he touched his patients one by one with the wand. Results varied considerably, the more suggestible the patient, the better the outcome. Some remained unaffected; others fainted or experienced strange sensations akin to insects crawling all over their bodies.

Mesmer believed that in order to cure someone he needed to provoke a 'crisis' within them. He thought that these sensations were the manifestation of that crisis, and declared a cure imminent.

Royally Exposed

Whether motivated by genuine concern or professional jealousy, the medical community continued to denounce Mesmer as a quack. Someone with royal influence must have had a quiet word though because, in

1784, King Louis XVI set up a royal commission to examine Mesmer's theories.

The king appointed four members of the Faculty of Medicine, including such eminent scientists as Benjamin Franklin and Joseph Guillotine. They stopped short of labelling Mesmer a charlatan, but were nevertheless unable to find any evidence to support Mesmer's belief in animal magnetism. Instead they attributed his results to 'a vivid imagination'. Their report stated;

Nothing proves the existence of magnetic animal fluid; imagination without magnetism may produce convulsions. Magnetism without imagination produces nothing.

Just as Mesmer was right for all the wrong reasons, the members of the Royal Commission were wrong for all the right ones. Far from a placebo effect, it was actually Mesmer's visualization and relaxation techniques that were yielding the real results. Interestingly the report did little to dent Mesmer's popularity with his patients, who obviously felt that the proof of the pudding was in the eating.

In scientific spheres, his reputation was even worse than it had been before the Royal court's intervention, and he faded swiftly into obscurity. 'Mesmerism', as it had begun to be known, was relegated to carnival side shows and music hall acts in place of treatment rooms.

It would be some considerable time before any benefits from Mesmerism would be realized. But eventually the scientists caught up with Franz Mesmer's thinking. The dramatic master of animal magnetism had been ahead of the game. Mesmerism and the healing hands of its leading light had all along been paving the way for the global development of hypnosis and hypnotherapy. Franz Mesmer's true legacy to the world had finally been revealed.

Caricature of Franz Mesmer (1779) depicting him as an ass hypnotizing a female subject with a finger. Considered by many to be a quack and charlatan, Mesmer's ideas of animal magnetism were widely ridiculed.

THE LOBOTOMISTS

Egaz Moniz (1874 – 1955). An ambitious, multitalented neurologist and politician, history has judged Egaz Moniz harshly. True, when he began developing the surgical procedure known as pre-frontal lobotomy, he had no clinical psychiatric experience and no real interest in psychology, but the problem of mental illness was one that no one had a solution to, and some people just can't resist a challenge like that.

Antonio Egaz Moniz was born in 1874 in Avanca, Portugal, to Fernando de Pina Rezende Abreu and Maria do Rosario de Almeida e Sousa. He was educated by his uncle before enrolling at the University of Coimbra to study medicine. He later studied at Bordeaux and Paris before heading back to Coimbra where he became a professor in 1902. In 1911 he transferred to the Chair of Neurology in Lisbon, and it was here that he worked for the rest of his life.

Moniz was inspired by the work of Gottlieb Burckhardt, a Swiss psychiatrist who performed the first psychosurgery on a human in the late 1880s. Burckhardt operated on six patients at Prefargier Asylum, of which he was head, removing a piece of the cerebral cortex. One of the patients died a few days after the procedure, but, in general, Burckhardt concluded that these experiments were a success and presented his findings to the Berlin Medical Congress. His fellow surgeons were less than impressed. In fact, he received such a poor reception that

Burckhardt ended his research on the topic and performed no more operations of the kind.

The idea of tampering with the personality using a scalpel is abhorrent, but Burckhardt and Moniz developed their procedures at a time when the vast majority of treatments for mental illness were at best ill-informed, and, at worst, tantamount to torture. The leucotomy was just one of a series of very invasive procedures introduced in Europe during the first part of the 20th century, including malarial therapy, barbiturate-induced deep sleep therapy, insulin shock therapy, cardiazol shock therapy and electroconvulsive therapy. All of which frequently caused drastic and life-changing side effects – some were fatal.

The scientific community was so desperate to understand the workings of the human brain better, and to offer treatment for mental illness, that it was willing to consider trying anything at all.

MRI scans of the brain showing the white cerebral matter.

Cutting the White Cerebral Matter

Moniz was regarded by many as an eminent neurologist thanks to his work on cerebral angiography, which developed a way to visualize the blood vessels in the brain. He named the leucotomy after the Greek words for 'cutting white' – referring to the brain's white matter.

The procedure, which was performed by the surgeon Pedro Almeida Lima under Moniz' direction, involved drilling holes in the patient's skull and destroying tissue in the frontal lobes by injecting alcohol into them. By doing so, Moniz hypothesized that he would destroy the neural pathways that had created unhealthy emotional loops in the patient's brain. As the brain healed itself, new pathways would be formed, and with them, new, more positive behaviour would emerge.

Later, Moniz changed the technique, discarding the alcohol injections in favour of an instrument he had developed called a

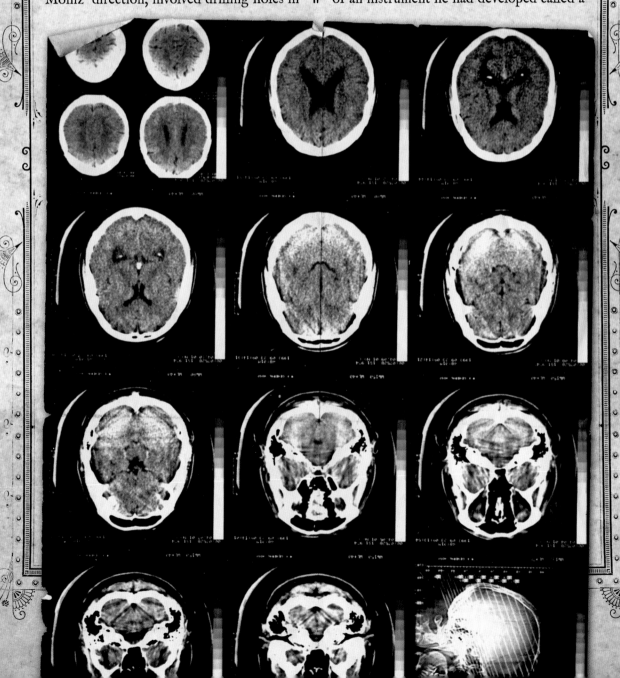

leucotome, a narrow-shafted implement that was inserted into the drilled holes, before cutting the brain tissue with a retractable wire loop.

Curing Madness

Between November 1935 and February 1936, Moniz and Lima performed the leucotomy procedure on 20 patients; 12 females and eight males, all between the ages of 27 and 65. Of those patients, nine were suffering with depression, six with schizophrenia, two from panic disorders and one each from manic depression, catatonia and ma-

nia. By their own assessment, 35 per cent improved dramatically, 35 per cent showed mild improvement and the other 30 per cent displayed no change whatsoever.

Certainly, the side-effects were nothing like as dramatic as those seen during the course of Burckhardt's research. All the patients survived, and so it was deemed a success. Crucially though, Moniz and Lima failed to perform any analysis of the long term effects of the leucotomy, choosing to evaluate the patients' progress after a period of weeks, rather than months, or indeed, years.

Dr Walter Freeman performing a lobotomy in 1949 using an instrument like an ice pick which he invented for the procedure. The instrument was inserted under the upper eyelid of the patient to sever nerve connections in the front part of the brain.

Nevertheless, in 1949 Moniz became the first Portuguese national to be awarded the Nobel Prize for his 'discovery of the therapeutic value of leucotomy in certain psychoses'. Moniz was very cautious in his promotion of the leucotomy. He recommended it only for very specific types of mental illness, and only when everything else had been tried. Others were not so discriminating.

Pychosurgeons Rack their Brains

Among the men and women who helped undermine the whole notion of psychosurgery were Dr Walter Freeman and Dr James Watts. Freeman was fascinated by Moniz' work in the field. He and Watts performed the first prefrontal lobotomy on American soil in 1936 at the Hospital of George Washington University in Washington.

The patient was 63-year-old Alice Hood Hammatt, a housewife from Topeka, Kansas, who suffered from agitated depression, symptoms of which included stomach ulcers, insomnia and anxiety. Emotional problems had plagued her for most of her life and she'd suffered numerous nervous breakdowns. Without a viable alternative, Mrs Hammatt faced permanent incarceration in a mental hospital. The surgical alternative provided by Freeman may have seemed extreme, but given the situation, the Hammatts felt they had little real choice. So they gave Freeman and Watts their consent.

Freeman and Watts had already ordered several of Moniz' leucotome instruments from Paris, and had begun to practise cutting the nerve fibres of the prefrontal lobes in preserved human brains. Watts seems to have been well aware of the risks associated with the procedure, but the bulk of his concern lay with his career, rather than with the life of their patient. He knew that by conducting such a controversial procedure, he might be ostracized and labelled a 'kook' by the medical fraternity. Conversely, Freeman, the son of a famously innovative surgeon, quite liked the idea of being a radical.

Alice in Wonderland

After struggling to anaesthetize a highly anxious Alice, the two surgeons followed closely the operation described by Moniz. The whole procedure lasted about an hour and was, by Freeman, Watts, and the Hammatt family, deemed a remarkable success.

Less than three weeks after the surgery, Freeman and Watts presented their findings to the District of Columbia Medical Society, claiming to have relieved Alice Hammatt of her symptoms. Within weeks, their paper on the use of prefrontal lobotomy to treat agitated depression appeared in the Medical Annals of the District of Columbia. As Watts had suspected, their claims met with a cynical response but Hammatt reported a distinct change in her anxiety levels and her husband called the five years following the operation the happiest of his wife's life.

Alice struggled to write legibly and at times appeared 'a little too placid'. She was unable to initiate conversation or take physical actions of her own accord, but Freeman and Watts declared that the benefits of the procedure far outweighed its detrimental side-effects.

Fast Track Lobotomy

Bolstered by this early success, Freeman and Watts set about assimilating the lobotomy into mainstream psychiatric medicine. They went on the campaign trail, promoting the use of lobotomy to treat everything from depression to criminality. In the United States alone, 40,000 patients received lobotomies over the next 40 years. Freeman himself conducted up to 3,500 of these operations.

Frustrated by what he saw as unnecessarily fiddly procedures and long operating times, he began designing a quicker way to achieve the same effect, something that could be rolled-out to the mass market of the overcrowded mental institutions and correctional facilities where budgets were tight and basic equipment, such as general anaesthetic, was scarce.

The ice-pick lobotomy required only local anaesthetic, or electroshock treatment if anaesthetic could not be obtained, and could be completed in as little as 10 minutes. Freeman apparently had his assistants time him in a bid to improve his personal record. The procedure involved inserting an ice pick shaped implement above each of the patient's eye sockets, driving it through the thin bone with the help of a mallet, moving the ice pick back and forth.

Howard Dully

The relative ease with which a lobotomy could be obtained and performed helped to fool people into believing it was a 'cure-all' for mental illness, and soon even the parents of unruly children were coming forward, begging Freeman to 'fix' their offspring.

One such set of parents were those of Howard Dully, a 12-year-old boy whose uncooperative behaviour was troubling his stepmother, Lou. According to Freeman's notes she said he was refusing to go to bed, he was 'defiant' and 'savage looking'. If that sounds to you like the manner of a perfectly normal 12-year-old, you'd be correct.

There was nothing psychologically wrong with Howard. Nothing, at least, that wouldn't be fixed with the passage of time. But by this time Freeman had decided that his ice pick lobotomies were suitable not just for the incurably mentally ill, but for anyone that failed to comply with society's norms. Freeman's notes of December 3 1960 read; 'Mr and Mrs Dully have decided to have their son operated on. I suggested they did not tell Howard anything about it.'

When Freeman met with Howard a fortnight after his lobotomy, he told Howard what had been done to him, 'he took it without a quiver. He sits quietly, grinning most of the time and offering nothing.' Ultimately Howard was made a ward of the state.

Mr and Mrs Dully chose not to ever discuss the procedure with Howard or explain to him why they had given Freeman their consent. As a consequence, Howard spent years trying to put the pieces of his personality back together, always feeling 'different', as though there was a part of his soul missing. But Howard Dully was not alone in losing more than he gained.

The Sweet Smiling Princess

Another famous recipient of Freeman's lobotomy was Rosemary Kennedy, the younger sister of US President John Kennedy. It was widely believed, but never officially confirmed, that Rosemary had learning difficulties. She went to school but was taught separately from the other students. She certainly did not live up to the high standards expected of a member of the Kennedy family but the precise reasons for her perceived 'difference', remains unclear. Was she mentally ill, or retarded? Why did the Kennedy family go to such great lengths to conceal her from the world? Why precisely did they decide to have her brain operated on? We shall never know for sure.

Whatever the exact circumstances, Rosemary's situation became increasingly difficult to manage as she began to take an interest in boys. The interest was reciprocated by young men who mistook Rosemary's behaviour (whatever it was) for naivety and feminine mystique. Rosemary's biographer describes her as a 'snow princess with flush cheeks, gleaming smile, plump figure and a sweetly ingratiating manner to almost everyone she met.'

No wonder the Kennedys grew concerned when she began having mood swings and sneaking out at night. Doctors advised that a lobotomy would help to calm Rosemary down and so she was booked in for the op. The notes of James Watts describe what happened next.

We went through the top of the head. I think she was awake. She had a mild tranquillizer. I made a surgical incision in the brain through the skull. It was near the front. It was on both sides. We just made a small incision, no more than an inch.

As Dr Watts cut, Dr Freeman put questions to Rosemary, asking her to recite the Lord's Prayer and count backwards. When they stopped getting answers from Rosemary, they stopped cutting. Tragically, the procedure left Rosemary in a childlike state, with urinary incontinence and unintelligible speech. She required round-the-clock care for the rest of her life. Her bad temper had disappeared, but it had taken most of the rest of her with it.

Heading for Oblivion

Dr Freeman fell out of favour in America in 1967 when a repeat patient named Helen Mortensen died of a brain haemorrhage following her third lobotomy. It was to be his final operation of the kind. The hospital in question revoked his surgical privileges, and he retired soon afterwards. The procedure has since been characterized as 'one of the most barbaric mistakes ever perpetuated by mainstream medicine' and many countries have actually banned the practice, believing it to be contrary to basic human rights.

It is widely believed that the lobotomy has been completely abandoned by medical science, but where it remains legal, it is still used to treat a very small number of cases, to control chronic pain and to lessen the danger caused by traumatic brain injuries. These days it is generally accepted that the detrimental effects – childishness, apathy, irresponsibility and post-operative incontinence are the rule, rather than the exception. It is no longer regarded as a miracle cure for mental illness.

Cutting the Stone, also called *The Extraction of the Stone of Madness* or *The Cure of Folly*, by Hieronymus Bosch. Museo del Prado, Madrid. Completed around 1494.

The History of Trepanning: Human beings have been drilling holes in their heads for millennia. In fact, trepanning – the practice of removing bones from the skull and exposing the outer-layer of the brain – may be the oldest form of surgery known to man. Evidence of its use has been found in human remains dating from the Neolithic period onwards. Cave paintings indicate that trepanning was used to treat a plethora of complaints including epileptic seizures, migraines and mental disorders. Evidence also suggests that trepanning was used as emergency surgery to treat head wounds typically caused by slings and war clubs. Of 120 skulls found at a burial site in France dating back to 6500 BC, one third displayed signs of trepanation. Many of the skulls also showed signs of healing, leading archaeologists to conclude that the patients must have survived the surgery. So, when the Portuguese neurologist Egaz Moniz began to devise an invasive procedure known as leucotomy in the mid-1930s, it was by no means a new idea.

MIND GAMES

José Delgado (1915 – 2011).   The impassioned prophet of a new 'psycho-civilized society' whose members could alter their own mental functions, José Delgado set out to provide a less destructive alternative to the lobotomy. But the shocking applications of his mind control work were wide open to evil manipulation by malevolent forces.

José Manuel Rodriguez Delgado was born in 1915, in Ronda, Spain, the son of a Spanish eye doctor. He earned a medical degree from the University of Madrid in the 1930s and intended to follow in his father's footsteps, but a stint in a physiology laboratory, plus exposure to the work of the Spanish neuroscientist Santiago Ramón y Cajal put paid to all that. He was suddenly fascinated by the mysteries of the brain, and particularly intrigued by the work of the Swiss physiologist Walter Rudolf Hess, who had demonstrated that he could produce certain basic behaviours, such as rage, hunger and sleepiness in cats by electrically stimulating their brains.

The Stimoceiver

In 1946, Delgado took up a year-long fellowship at Yale University, and four years later, he accepted a position in the Department of Physiology which was headed-up by the eminent psychiatrist John Fulton, a man whose work had also inspired Delgado's contemporary, the pioneer lobotomist, Egaz Moniz (see p. 84). At this juncture in medicine, the lobotomy was gaining popularity as a miracle treatment for mental illness. Delgado was disturbed by what he saw as an unnecessarily destructive, permanent and invasive procedure. He felt there must be way to offer a more conservative treatment by applying electrical stimulation directly to the brain.

Delgado's prototype 'stimoceiver' was an apparatus straight out of the props department of a horror movie set. It featured wires that ran from the implanted electrodes in the patient's brain out through the skull and skin to bulky electronic devices. These recorded the data but greatly restricted the patient's movements and left them prone to infection. As a way to get round this problem, Delgado set about designing a much smaller device that could be implanted in the brain in its entirety and controlled using radio waves.

Fighting Bulls and Spanish Guitars

Delgado's early experiments involved cats, monkeys, chimpanzees, gibbons and, perhaps most famously, fighting bulls. Later, he began experimenting on humans. Between 1952 and 1974, Delgado implanted electrodes in 25 people at a now defunct mental hospital in Rhode Island.

The majority of his patients suffered from schizophrenia or epilepsy. All were, according to him, very ill patients whose disorders had resisted all other available treatment. Unlike contemporaries like Freeman and Watts, Delgado turned away many more patients than he treated. He was inundated with requests from concerned parents asking him to fix their damaged and vulnerable offspring.

A young woman was brought to Delgado to find a cure for her promiscuity. In the absence of evidence suggesting a confirmed neurological disorder, he refused to operate, stating that the technology was still too primitive to treat anything but the most desperate and the most clear cut cases.

In human patients, Delgado showed that, by stimulating the motor cortex, he could control his subject's movements. Even when they tried to resist, the electricity was stronger than the patient's will. By applying electricity to various parts of the limbic system he could also produce emotions, inducing fear, hilarity, rage and even lust.

Some reactions were so dramatic that they startled the researchers conducting the tests. In one experiment Delgado and two collaborators stimulated the temporal

Delgado animal experiment, 1963. These monkeys with electrodes implanted in their brains and battery backpacks certainly look like relatives of the Frankenstein monster. Early research such as this paved the way for more recent ground-breaking developments in the treatment of neurological diseases.

lobe of a 21-year-old epileptic woman as she sat calmly strumming a Spanish guitar. She flew into a rage and smashed the guitar to smithereens.

Delgado the Matador

Delgado and his colleagues found that they could use the technique to stimulate euphoria intense enough to counteract depression and even physical pain. However, the treatment was unreliable, effects varied greatly from person to person and the same patient could react differently to the same stimulus when the experiments were repeated on another day.

In 1963, José Delgado performed the most famous experiment of his career when he inserted stimoceivers into the brains of several fighting bulls at a bull-breeding ranch in Cordoba, Spain. One-by-one each bull was led into the bullring where Delgado stood in the traditional role of matador. Using a remote control, Delgado was able to control the bulls' physical responses.

In one instance he managed to stimulate the caudate nucleus in order to force a 1000 lb charging bull to skid to a halt yards away from him. For Delgado, the experiment was a scientific success and massive PR triumph. The *New York Times* called the event 'the most spectacular demonstration ever performed of the deliberate modification of animal behaviour through external control of the brain.'

Delgado may have appeared confident in front of the press, but he later admitted feeling frightened just before forcing the bull to change his course. The experiment also struck fear into the hearts of many ordinary Americans. If an enormous and powerful beast like a bull could be manipulated by remote control, what else could be achieved? Could humans be as easily controlled if the device fell into the wrong hands?

Building a Race of Slave Soldiers

Many of Delgado's colleagues felt there was an undeniably ominous side to his research and they worried about where it might lead. It is true that Delgado had sponsors within the Pentagon, but he insists (perhaps naively) that they viewed his work as little more than basic research, and took no interest in its military applications. He called himself a pacifist and always dismissed the idea that the technology he developed could ever be used to build a race of slave soldiers. While brain stimulation can increase or decrease aggressive behaviour, it cannot determine the target of such aggression.

With exposure to enough carefully targeted propaganda, this 'technical hitch' could be overcome so that a soldier's aggression could be aimed at any social, religious or ethnic group. However, history has shown us that the world's soldiers are generally capable enough of killing on command without the help of brain implants. The brain chips Delgado developed would not have been capable of making a soldier physically stronger, more intelligent or more able to kill, since they could only impact upon existing brain signals – in other words they couldn't create knowledge, or memories – but they could change feelings. For many, that was power enough.

The Violence of the Brain

Frank Ervin and Vernon Mark were two researchers at Harvard who briefly collaborated with Delgado. In their controversial book, *Violence and the Brain*, they suggested that psychosurgery might help to quash the aggression displayed by black rioters in the American inner cities. As a result all their funding was withdrawn.

Around the same time another psychiatrist claimed that he had been able to change the sexual orientation of a homosexual male by stimulating his septal region whilst he had sexual intercourse with a female prostitute.

This had little to do with Delgado personally, but it threw a negative light on his work and played into the hands of people who argued that evil applications of such technology outweighed the potential benefits to human kind.

The respected psychiatrist, Peter Breggin, accused them of trying to create a society in which anyone who differed from the norm, faced surgical mutilation. The same accusation that was levelled at the lobotomists, the very people Delgado strived to disassociate himself from.

The Anxiety of Paddy the Chimpanzee

The bullring experiment may have been his most famous to date, but Delgado himself did not believe it to be the most scientifically significant. That accolade goes to an experiment he conducted on a female chimpanzee called Paddy, who displayed symptoms of an anxiety disorder. Delgado programmed Paddy's stimoceiver to pick up distinctive brain signals called spindles, which are spontaneously emitted by the brain's amygdala.

When Paddy's stimoceiver detected a spindle it produced a painful or unpleasant sensation. After two hours, the unfortunate animal's amygdala produced 50 per cent fewer spindles. Within six days, the frequency had dropped by a massive 99 per cent.

Unsurprisingly, Paddy was left somewhat traumatized by her experience with Delgado, but the neurosurgeon felt he had made an important breakthrough.

Although this monkey experiment from the late 1960s may represent everything evil-looking about animal research, Delgado's electrical techniques are still in use today and are now helping to treat various neurological pathologies. Deep brain stimulation (DBS) is a surgical treatment involving the implantation of a medical device called a brain pacemaker, which sends electrical impulses to specific parts of the brain. DBS is a promising therapy for epilepsy, depression, Alzheimer's and stroke.

SHOCK TACTICS

Stanley Milgram (1933 – 1984). The Nazi Holocaust is widely regarded as the most tragic and appalling episode in the history of humankind. Given its moral reprehensibility, did all the participants share a perverted sense that they were doing the right thing, or were they simply doing as they were told by people in authority? In 1961, Stanley Milgram set out to find an answer to this question. The Milgram Obedience Experiment has since become the most famous experiment in the history of social psychology. Its findings are as conclusive as they are, for want of a better word … shocking.

Stanley Milgram was born in New York in 1933, to an Eastern European Jewish couple who owned a bakery and earned a modest income. He excelled throughout his schooling and studied political science at Queens College before deciding to switch paths and focus on psychology instead. After some catch-up studying (he'd never studied psychology before) he was eventually accepted by Harvard University where he earned a PhD in social psychology in 1960.

Milgram's parents were holocaust survivors, so perhaps the motivation for his obedience experiment came from personal questions that arose from their experiences. His controversial methods definitely point to something of a maverick character, and certainly to someone who was motivated by more than just scientific endeavour.

Your Obedient Servant

First, Milgram took out an advert in a local newspaper offering participants US$4.50 for one hour's work as part of a psychology study at Yale University, the purpose of which was to test memory and learning. Successful participants were told that the specific remit of the study was to investigate the links between punishment and study. All the participants in the original experiment were men, and all were ordinary local citizens.

They were split into pairs and randomly assigned the role of teacher or learner. The learner was then strapped into a chair and electrodes attached to their arm. The electrodes were plugged into an electric shock generator, it was then explained to the pair that the shocks would serve as punishment for any wrong answers given during the test that was about to commence. The teacher

Public Announcement

WE WILL PAY YOU $4.00 FOR ONE HOUR OF YOUR TIME

Persons Needed for a Study of Memory

• We will pay five hundred New Haven men to help us complete a scientific study of memory and learning. The study is being done at Yale University.

• Each person who participates will be paid $4.00 (plus 50c carfare) for approximately 1 hour's time. We need you for only one hour: there are no further obligations. You may choose the time you would like to come (evenings, weekdays, or weekends).

• No special training, education, or experience is needed. We want:

Factory workers	Businessmen	Construction workers
City employees	Clerks	Salespeople
Laborers	Professional people	White-collar workers
Barbers	Telephone workers	Others

All persons must be between the ages of 20 and 50. High school and college students cannot be used.

• If you meet these qualifications, fill out the coupon below and mail it now to Professor Stanley Milgram, Department of Psychology, Yale University, New Haven. You will be notified later of the specific time and place of the study. We reserve the right to decline any application.

• You will be paid $4.00 (plus 50c carfare) as soon as you arrive at the laboratory.

TO:
PROF. STANLEY MILGRAM, DEPARTMENT OF PSYCHOLOGY, YALE UNIVERSITY, NEW HAVEN, CONN. I want to take part in this study of memory and learning. I am between the ages of 20 and 50. I will be paid $4.00 (plus 50c carfare) if I participate.

NAME (Please Print)..

ADDRESS ....

TELEPHONE NO. .... Best time to call you ....

AGE........ OCCUPATION.................... SEX......
CAN YOU COME:

WEEKDAYS .... EVENINGS ......WEEKENDS........

would be given a number of word pairs to read aloud to the learner, all the learner had to do was repeat them back.

The experimenter assured both the teacher and the learner that, although they would be painful, the shocks would cause no permanent tissue damage. The pair were then taken into separate rooms, divided by a thin wall. The teacher sat at a machine with 30 switches, each labelled with terms including 'slight shock', 'moderate shock', 'danger: severe shock'. The final switch was labelled simply 'XXX'. This would no doubt have been a very difficult and tense moment for both the teacher and the learner, except for the fact that the learner was not a learner at all, but a trained actor who was in on the whole thing.

The Enjoyment of Inflicting Pain

The purpose of the experiment was not to study learning at all, but to determine whether or not the test subject (the teacher) would willingly inflict an electric shock on a likeable stranger who, as far as they were aware, was randomly chosen and had done nothing to deserve such a brutal punishment. The actor's job was to answer the questions in a set way, and fake cries of excruciating pain as the shocks increased in intensity.

The research team were given set responses to give whenever the teacher raised concerns about the morality of what he was doing. These were designed to give the impression that they should continue despite the learner's obvious distress.

The first time the test subject complained, the researcher would simply say 'please continue'. The second time: 'The experiment requires that you continue.' The third: 'It is absolutely essential that you continue.' And the fourth time: 'You have no other choice. You must go on.'

If, after their fourth appeal, the test subject was still unwilling to pull the switch, the experiment was concluded and the truth revealed to the distressed participant. If the test subject stopped complaining and pulled the switch, the experiment would continue until he had inflicted three shocks at the

Milgram Experiment, 1963. A poster asking for volunteers for Stanley Milgram's obedience to authority experiment, euphemistically called A Study of Memory in the poster.

maximum voltage, a massive 450 volts.

Interestingly, some of the teachers made attempts to excuse their behaviour, saying things like; 'This guy is so dumb he deserves to get shocked.' This mirrors the process of dehumanization in war, or wherever one group of people is actively oppressed by another.

The Shocking Results

Before conducting his behavioural experiment Stanley Milgram asked 14 Yale University senior-year psychology majors to predict the behaviour of 100 hypothetical test subjects. They all believed that a very small percentage (between 1 and 3 per cent) would willingly inflict the maximum voltage on the learner. They were wrong. In fact 65 per cent were willing to inflict the most severe shock. 100 per cent of the test subjects were compliant up to 300 volts or more.

Milgram repeated his obedience experiment at least 19 times, varying factors such as the gender of the test subject, the immediacy of the 'learner' and the proximity of the experimenter.

Although female test subjects reported more stress than their male counterparts, they were just as likely to deliver the maximum shock. There were no differences between the sexes.

The Closer I Get to You

The immediacy of the learner had a major impact on the results. When the learner was placed closer to the test subject, the test subject was less likely to comply with the experimenter's request. Similarly, the more distanced the experimenter was from the test subject, the less likely he was to obey.

When the experimenter communicated with the test subject by telephone rather than in person, a small number of test subjects pretended to continue with the experiment, rather than inflict pain on their learner. In another experiment, the teacher was accompanied by two other teachers (both, in fact, actors). Their presence and behaviour also had a major impact on the compliance of the test subject.

Slightly differing versions of Milgram's experiment were repeated by others in the 35 years following the original, and although different parts of the world reported slightly differing rates of obedience, there was no change over time, suggesting that people have not become more, or less, obedient over that period.

The Hofling Hospital Experiment

In 1966 the psychiatrist Charles K. Hofling decided to take the basic idea behind Milgram's experiment and test it in the field, focusing on the relationship between female nurses and male physicians. In a natural hospital nurses were telephoned by an unfamiliar doctor and ordered to administer what could have been a fatal overdose of a (fictional) drug to a patient under her care. During this short telephone conversation he told her he would sign for the drug later. The drug had been placed in the medical supplies cabinet but was not on the approved drug list. The bottle clearly stated that 10 mg was the highest safe dose. The doctor asked the nurse to administer 20 mg.

Many wondered after the horrors of WWII how people could be motivated to commit acts of such brutality towards each other. Stanley Milgram (left) was influenced by the events of the Nazi holocaust to carry out an experiment that would demonstrate the relationship between obedience and authority.

Before conducting the experiment Hofling asked a group of nurses and nursing students what they would do. The vast majority of the nurses and all the students said that under those conditions they would not administer the drug.

There were plenty of valid reasons why the nurses should not have agreed to administer that drug. Firstly, the label on the drug clearly warned that 10 mg was the recommended dose. Secondly, the drug was not on the approved drugs list and thirdly (and perhaps most importantly) it was utterly against hospital protocol for nurses to follow the instructions of doctors who were unknown to them, especially over the phone.

Out of the 22 nurses Hofling tested, 21 said they would have administered the drug. Only one nurse questioned her orders. Luckily, the other 21 were stopped at the door of the patient's room before they could give the drug.

The Real Deal

Charles Sheridan and Richard King believed that some of Milgram's original participants had realized that the shocks were fake, and had played along so as not to ruin the experiment. In 1972, they set up an identical experiment to Milgram's, with one key difference; there was a real victim, a puppy that was given real electric shocks.

Out of 26 participants, 20 obeyed to the end, 13 of those were women. Overall, 100 per cent of the women and 54 per cent of the men were fully compliant. Just as in Milgram's experiment, the women reported a great deal more stress than their male counterparts, but it didn't stop them doing the deed.

Ultimately, the results of Sheridan and Kings' experiment disproved their own hypothesis. As it turns out, given the right circumstances most people really will just do as they're told, even if it involves murder.

THE PSYCHOLOGY OF IMPRISONMENT

Philip G Zimbardo (1933 –). What happens when good people find themselves in an evil place? Will they continue to be good, or will the pressure of the situation prompt them to behave in unacceptable, or even evil ways? The Stanford Prison Experiment was set up by Philip Zimbardo to find answers to this very question. It was supposed to last for two weeks, but it did such psychological damage to the participants, that it had to be halted early. Goodness, it seemed, was dependent on more than just upbringing.

Philip Zimbardo was born in the South Bronx, New York City in 1933; the grandson of poor and uneducated Sicilian immigrants, whose parents never went to high-school. He was beaten up and discriminated against on a daily basis, but rather than convert him to a life of crime, it somehow taught him that education was the most effective way to escape poverty, and so he pursued it doggedly. He became the first person in his family to complete high school and graduated with a BA from Brooklyn College, where he triple majored in psychology, sociology and anthropology. From there, he went to Yale to study for his MS and finally, for his PhD, both in Psychology. He taught briefly at Yale University, Columbia University and New York University before becoming a faculty member at Stanford University in 1968. Given Zimbardo's background, it is perhaps unsurprising that he chose to focus his research on social environments and their impact on human behaviour.

When Good Kids Go Bad

The Stanford Prison experiment has become one of the most famous experiments in the history of psychology. It was funded by the US Office of Naval Research who wanted to determine the cause of conflict between military guards and detainees, and although such a study would be completely illegal under US law today, its findings have proved invaluable in helping us to understand how human rights abuses so often take place wherever one group of people is assigned control over another.

Zimbardo was a former high-school classmate of Stanley Milgram, and he was interested in expanding on the research undertaken by him, looking at deindividualization and the impact of social variables on human behaviour. The premise for the experiment went like this: 'suppose you had only kids who were normally healthy, psychologically and physically, and they knew they would be going into a prison-like environment and that some of their civil rights would be sacrificed. Put in that bad, evil place – would their goodness triumph?' *The line between good and evil is permeable and almost anyone can be induced to cross it when pressured by situational forces.* Philip G Zimbardo

Selection Procedure

Zimbardo and his team placed advertisements in the local newspaper, looking for male student volunteers to take part in a psychological experiment in return for US$15 a day (the equivalent of about US$80 a day in today's money). By the end of the rigorous selection process 24 predominantly white candidates remained, all of whom were American or Canadian, middle class, healthy and intelligent males. Next, they were randomly split into two equal-sized groups. Half would assume the roles of prison guards, and the other half, detainees.

The prisoners would be required to stay in the mock prison for the entire duration of the experiment, just as if they really had been found guilty of a crime. The guards were required to work shifts, with three guards to each shift. After each shift the guards were allowed to return home, just as if they were working in a real prison.

In an attempt to make the prison experience as realistic as possible Zimbardo also called upon a man who had spent 17 years behind bars, and had extensive experience of the prison system, to act as a consultant to the scheme. He had introduced the scientists to a number of ex-cons and prison staff during an earlier Stanford class they had taught called 'The Psychology of Imprisonment.'

The research team also had the support of the Palo Alto police force. Zimbardo assumed the role of prison superintendent within this fictitious scenario. It was a role he would come to embrace just as wholeheartedly as the other participants in the study.

The Bogus Basement Slammer

The prison was constructed in the basement of Stanford's Psychology Department. An open corridor was blocked off at either end to serve as the 'Yard', where the prisoners could walk, eat or exercise. They would be able to use the toilet down the hallway but would have to be blindfolded in order to do so, so as not to disclose the way out of the prison. Three, 6ft x 9ft prison cells were created by taking the doors of some of the laboratory rooms and replacing them with specially made doors with steel bars and cell numbers.

On one side of the corridor was a cupboard that became known as 'the hole', it was 2ft wide and 2ft deep – just big enough to fit a badly behaved prisoner. As it happened, this would come in useful on a number of occasions.

An intercom system allowed scientists to discreetly bug the prisoner's conversations and make announcements to the prisoners. All clocks were removed, and there were no windows on the block – adding to the prison feel. At one end of the hall was a small opening from which the participants were filmed as they went about their day-to-day lives.

Heavy Ankle Chains and No Underwear

Early in the morning, on day 1 of the experiment, the prisoners were arrested in their homes by real police officers, charged with armed robbery and processed at the local police station. A key feature of the study, often overlooked in analysis of it, was the mock arrests of those destined to be prisoners by the Palo Alto Police in their squad car. Each of those nine students were then booked at the police department prior to being brought blind folded to the Stanford Jail. Because the Authorities took away their freedom, only the authorities had the power to give it back—in the form of a parole. Had the prisoners voluntarily given up their freedom, Zimbardo knew that when the going got rough, they might elect to quit the experiment; but you can't quit a prison. They were then brought one by one to the jail and introduced to the prison warden (portrayed by an undergraduate research assistant of Zimbardo, named David Jaffe). As part of this checking-in process each prisoner was stripped of all clothing, intimately searched and deloused with the expressed intention of inducing as much humiliation as possible.

Each prisoner was issued, not with a standard prison uniform, but with a short dress bearing his prison ID number. From now until the end of the experiment, each prisoner would only be addressed by his number. Underclothes were not permitted. Each prisoner was forced to wear a heavy chain on his right ankle, which was bolted on and worn at all times. The only permitted shoes were rubber sandals, and on their heads they wore stocking caps made from women's hosiery—to reduce individuality of hair styles, short of cutting off their hair, as in real prisons and the military.

This was a functional simulation of a prison, not an actual prison. The test team wanted to create the same emotions and situations that take place during prison life, and for this reason they had to humiliate, dehumanize and institutionalize the prisoners as quickly as possible, acclimatizing them to their roles as prisoners. The research team wanted to make the prisoners feel emasculated and vulnerable as quickly as possible. They succeeded. The garb they were made to wear had the effect of making them walk and hold themselves more like women than men.

Turmoil and Retribution

The guards were given no specific training in how to be guards. They were simply told to do whatever they felt was necessary to maintain order within the prison and, perhaps crucially, to command respect from the prisoners. They made up their own set of rules, in a process that was overseen by the warden. All the guards wore identical khaki uniforms, all carried a whistle and a truncheon borrowed from the police. They all wore mirrored sunglasses (an idea borrowed from the film *Cool Hand Luke*) which prevented anyone from seeing their eyes or accurately reading their emotions.

It didn't take long for things to get interesting. On day 2 of the experiment, the prisoners in cell 1 blockaded their door, removed their stocking caps and refused to do anything the guards ordered them to do. The guards, warden and superintendent were taken aback by this turn of events, given how straightforward day one had been. The guards were left to decide for themselves what to do next.

First, they called for back up, and guards who were not on shift came in to provide reinforcements. They broke into the cell in question using fire extinguishers to blast the prisoners away from the cell door, then they stripped the rebels naked, took the beds out and forced the ringleaders into the hole.

Underprivileged Inmates

They decided to set up a privilege cell, and moved the inmates who had not been involved in the rebellion into this space, where they were given a good meal instead of bland prison food. The prisoners in the privilege cell chose not to accept the treat in order to avoid getting on the wrong side of the rebels.

The guards quickly became frustrated, enraged, even, by the prisoners attempts to assert their individuality, and, as the experiment progressed they expressed their frustration in ever more creative and humiliating ways. Prisoners were subjected to 'counts' at all hours of the day and night, both to assert the guards' control over the prisoners' and

to reinforce their new identities as mere numbers rather than individuals.

Any mistakes made during the count resulted in harsh punishments, from rounds of press-ups to the withdrawal of mattresses and bathroom privileges. Soon prisoners were being forced to defecate and urinate in buckets that were left in the corners of cells. The prison began to look, feel and smell, more like a real prison and the prisoners began to act less like volunteers, and more like real prisoners.

Prisoner #8612

Less than two full days into the experiment, prisoner #8612 began to display signs of severe emotional disturbance. His thinking became disorganized and he began crying uncontrollably, begging to be released. What happened next shows how unhealthy the scenario had already become. Instead of releasing the man, who, after all, was only a volunteer, Zimbardo and his team were already so embroiled in their roles as prison board members that they assumed prisoner #8612 was 'putting it on' in order to fake his way out of jail.

The ex-con consultant was brought in to speak to the prisoner, and he chided him for being weak, offering him the chance to inform on his fellow captives in return for no further harassment. It was a pivotal moment in the study. Reality for the participants had shifted.

The plight of #8612 should have alerted Zimbardo's team to the real ethical problems they were facing, but instead, it made them more determined to control the prisoners' every move. Prisoner #8612 returned to his cell and delivered a chilling message to his fellow inmates. What they thought was a simple psychology experiment had become something more. This was, in effect, a real prison. He told them that they were unable to leave.

The Great Escape Attempt

One of the guards overheard the prisoners talking about an escape attempt. According to the rumours prisoner #8612, who had had to be released earlier on mental health grounds, was apparently rounding up some friends to help him spring the other prisoners out of Stanford. But instead of studying the escape rumour mill and allowing things to take their natural course, Zimbardo and his team stopped collecting data and reacted like real prison board members.

They set about improving security to prevent the escape – the organ grinders had become the monkeys. They'd come to believe their own fiction to such an extent that they, and their research, became eaten up by it. They had the prisoners moved to a 'safe location' in another part of the university and Zimbardo waited alone for the escape team, with the prison seemingly abandoned, doors off their hinges. He even considered luring back and recapturing prisoner #8612, an act that would have amounted to kidnap, since he was no longer a willing volunteer in the scheme.

The escape never materialized. The rumour was just a rumour and nothing more. As a result the guards were became even more brutal in their treatment of the prisoners. The punishments became more humiliating and more elaborate. People were forced to do things they would never willingly do in real life – like cleaning out toilet bowls with their bare hands.

Four prisoners broke down emotionally, a fifth even developed a psychosomatic rash all over his body when he discovered that his 'parole' had been turned down, others tried to do everything they could to win favour from the guards, which in turn, made them unpopular with the other inmates. There was no longer any group unity, just a bunch of isolated individuals hanging on, like in a concentration camp or mental hospital. But still Zimbardo was unwilling, or unable, to see that things had gone too far.

The Ultimate Dawn of Realization

By the fifth night, Zimbardo knew his experiment was in trouble. Not only had it become clear that some of the more aggressive guards were stepping up their abuse during the early hours of the morning, when they believed that no one would be filming them but the prisoners had begun to display pathological behaviour.

Christina Maslach was a recent Stanford PhD student who had been brought in to conduct interviews with the guards and prisoners. When she saw the prisoners being marched on a toilet run, legs chained together, bags over their heads, she became distressed and angry and she told Zimbardo exactly what she thought of his experiment.

Of 50 adults who had entered the mock jail during the course of the experiment, including a lawyer and a Catholic priest, and parents and friends on several visitor nights, she was the only one who openly questioned the morality of what Zimbardo and his colleagues were doing. Once she had, the spell was broken. By standing up and speaking out, this young woman acted heroically because she was aware of potential costs to her budding career and budding romantic relationship with Zimbardo. He was her main academic reference and new boyfriend. But she was willing to risk both losses to voice her distress over the inhumanity she was witnessing. Zimbardo was made aware that he too had fallen under the spell of the same powerful situational forces he created for the prisoners and guards: He had become the Superintendent of the Stanford Prison, not just the principal investigator of the Stanford Prison Experiment. It became clear to him then that the study should be ended, and so it was the next morning.

Only six days into a 14-day-long programme, the Stanford Prison Experiment was shut down. Later, when interviewed, one of Zimbardo's anonymous mock guards had this to say about his experience:

I really thought that I was incapable of this kind of behaviour. I was surprised, no I was dismayed to find out that I could act in a manner so absolutely unaccustomed to anything I would even really dream of doing. And while I was doing it I didn't feel any regret. I didn't feel any guilt. It was only afterwards when I began to reflect on what I had done, that this behaviour began to dawn on me and I realized that this was a part of me I hadn't really noticed before.

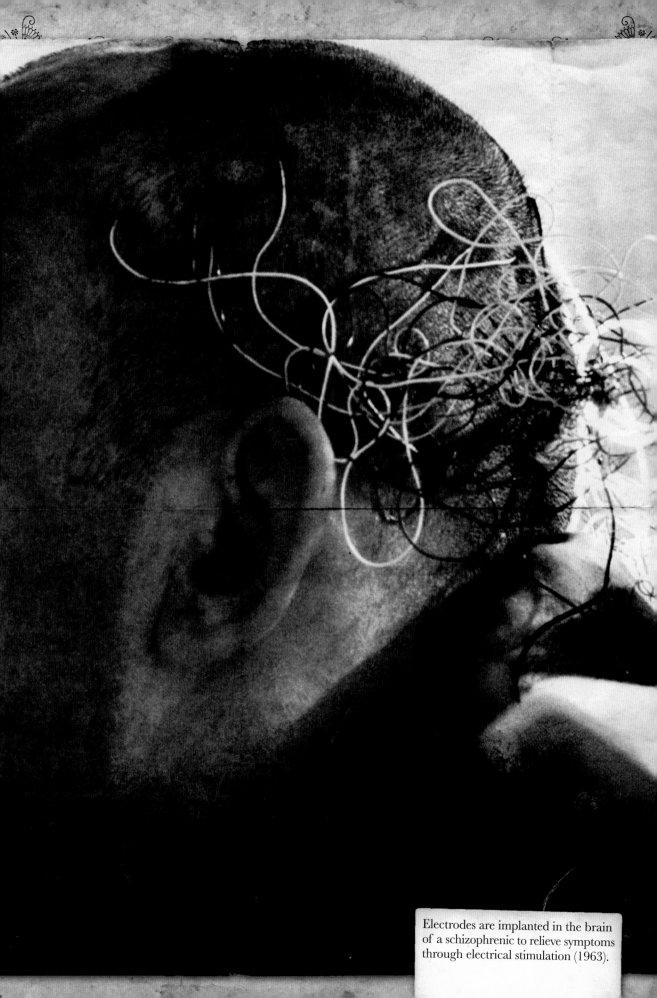

Electrodes are implanted in the brain of a schizophrenic to relieve symptoms through electrical stimulation (1963).

BRAIN-WASHING THE UNWITTING

Project MKULTRA. Whenever governmental organizations like the CIA are publicly accused of conducting secret experiments, cynics tend to yell 'conspiracy'. But do you seriously think that our secret services always operate within the law? Do you imagine for even one moment that they would pass up the chance to find out how LSD or torture affects soldiers for fear that it might compromise their human rights? Wake up, smell the coffee and step down off of that rainbow my naive friend. Welcome to the world of Project MKULTRA.

Project MKULTRA was the umbrella code name for a secret, illegal human experiment programme, which was run by the CIA from some point during the first half of the 20th century until it was closed down in the early 1970s. CIA Director Richard Helms ordered that all the MKULTRA files be destroyed in 1973, so what is known of MKULTRA has been gleaned from anecdotal evidence and from the handful of files that managed to escape the shredder, and were made public in 1977 under the Freedom of Information Act.

The purpose of MKULTRA was to investigate and manipulate individuals' minds, using hypnotherapy, sensory deprivation, isolation, torture, abuse and psychedelic drugs. In 1977, on the floor of the United States Senate, Senator Ted Kennedy, admitted as much, when he said:

The Deputy Director of the CIA revealed that over 30 universities and institutions were involved in an 'extensive human testing and experimentation' programme which included covert drug tests on unwitting citizens 'at all social levels, high and low, native Americans and foreign'. Several of these tests involved the administration of LSD to 'unwitting subjects in social situations'.

Operation Paperclip

In order to understand how a country that calls itself the 'Land of the Free' arrived at the idea of experimenting on its own citizens, it is important to look at the socio-political backdrop against which it was operating.

In the immediate aftermath of World War II, America was busy harvesting the secrets of the superpowers they had just conquered; the

Germans and the Japanese. Under the title of Project Paperclip, the CIA set about recruiting former Nazi scientists, some of whom had researched torture and mind control as part of their wartime role. Many of whom had been successfully prosecuted for war crimes during the Nuremberg Trials.

Paperclip branched off into a number of other highly classified research programmes, including Project Chatter and Project Blue-bird (later named Project Artichoke), the purpose of which was to take what the Nazis had learned about mind control, interrogation techniques and behavioural modification, and develop techniques to help fight the Soviets, China and, to a lesser extent, home grown Communists.

MKULTRA was a later addition to this group of programmes. Activated on 13 April 1953, it was headed up by the chemist Sidney Gottlieb, at the request of then CIA director Allen Welsh Dulles.

The Black Sorcerer

Sidney Gottlieb was born under the name Joseph Scheider, in the Bronx, New York City in 1918, the son of Hungarian Jewish immigrants. He graduated at the top of his class from the University of Wisconsin but was barred from participating in World War II because he had a clubfoot. Gottlieb's hobbies included folk dancing and goat herding. He did not sound like an ideal candidate for the Evil Scientist of the Year Award.

Gottlieb came to the CIA in 1951 as a poisons expert, with the typically vague job title of Chief of Technical Services. A frequent user of LSD himself, it was re-

ally only a matter of time before he began exploring its more sinister applications. One of Gottlieb's masterstrokes involved setting up a number of brothels in the San Francisco area, where prostitutes would routinely and surreptitiously drug their clients so that CIA agents could monitor the effects.

The idea being that the clients would be too embarrassed and ashamed to talk about their ordeal, and that, if they did, they wouldn't be believed. Shockingly, the CIA continued to run 'Operation Midnight Climax' for a total of eight years. The man Gottlieb put in charge of the project, George White, apparently described his career in this memo, which he sent to his superior at the Federal Bureau of Narcotics. The tone indicates that drug taking certainly was not limited to the guinea pigs themselves, but had become endemic in the culture of the CIA.

I was a very minor missionary, actually a heretic, but I toiled wholeheartedly in the vineyards because it was fun, fun, fun. Where else could a red-blooded American boy lie, kill, cheat, steal, rape, and pillage with the sanction and blessing of the All-Highest?

Exploiting the Population

But the abuse did not begin, or end with the activities of George White, or with CIA-controlled brothels. Gottlieb and his team exploited anyone they felt could not fight back – criminals, mental patients, prisoners and the very poor.

Bizarrely some people did actually volunteer to take part in the MKULTRA tests, and for these poor souls, they reserved the worst possible experiments. A group of seven Kentuckian volunteers were drugged with LSD for 77 consecutive days. In the middle of the experiment, when the participants had developed a considerable degree of tolerance to LSD, the experimenters began doubling and tripling the doses in order to 'break through this tolerance'. No one knows exactly what became of the men who took part in the test, but the long-term effects would have amounted to severe Post Traumatic Stress Disorder (PTSD).

The drug testing carried out by MKULTRA was ultimately responsible for at least one death and several others were driven totally and irretrievably mad. To add insult to injury, Gottlieb ultimately concluded that none of his work in this arena was of any practical use, but then, he would say that, wouldn't he? The alternative would mean admitting the unthinkable; that America adopted and used these tools in the field, which can't be true, obviously.

The Manchurian Candidate

In addition to studying the effects of psychedelic drugs on unsuspecting test subjects, another major aim of MKULTRA was to create better, stronger and more pliable soldiers. The Americans had heard about Soviet attempts to create a real life Manchurian Candidate, a soldier who can be brainwashed into doing whatever he is told, no matter how illegal, dangerous or immoral, and, as with the moon landing, they desperately wanted to get in first.

Using hypnosis, MKULTRA doctors like Dr George Estabrooks found that they could

even split personalities, so that an operative could go undercover on a mission without even knowing it. These agents gave away no tell-tale signs of deceit, and were not susceptible to torture because their conscious mind held no useful information. Secrets were all tucked safely away in their unconscious minds, and could only be unlocked by someone who had a special relationship with the agent, and knew exactly where to look.

Estabrooks performed this experiment on a suggestible Marine Lieutenant Jones, turning him into Jones A and Jones B in order to help him infiltrate the Communist party. Jones A became a completely different man when compared to his former self. He began to accept and to talk communist doctrine, he joined the Communist party and became a fully-fledged card carrying member. Jones A believed everything he said and thought as though he had always held those beliefs. It wasn't until he was under hypnosis that Estabrooks could access Jones B, a loyal and patriotic American who knew all about Jones A, and could relay the information he had collected.

Candy Jones: A Model Spy

One of the strangest stories concerning MKULTRA mind control experiments is that of Candy Jones, the World War II pin up who, years later confessed, under hypnosis, to living a second secret life as a CIA hypno courier during the Cold War.

Candy Jones was born Jessica Arline Wilcox in 1925. She was a leading fashion model, writer and radio talk show host who had taken part in a show called *Cover Girls Abroad*, and entertained troops in the Philippines at the end of World War II. While there she fell ill, and was treated by a doctor thought to be Dr William Kroger, a prominent psychologist and pioneer of hypnosis who was believed to be associated with UCLA.

According to Candy's taped interviews, she suffered an unhappy and abusive childhood, and had an imaginary friend named Arlene in order to help her through the tougher times. Dr Kroger (if it was him), seems to have accessed this information and manipulated this imaginary relationship to split Candy's personality into two halves.

The plan was that Candy would continue her life as a fashion model and business woman, while Arlene would carry out top-secret missions for the CIA. Later, in 1960, an old army acquaintance from her tour of duty in the Philippines approached her and asked to use her modelling school as an address for receiving letters and packages. Out of some vague sense of patriotic duty, Candy said yes.

Undercover Operations

Many loyal American citizens performed similar roles for the CIA during this period. Eventually she was asked to deliver a package to Oakland, California, and, when she went to deliver it, she was surprised to discover that the person on the receiving end was the same man who had treated her back in the Philippines. Candy agreed to be hypnotized by this man and, like a long forgotten telephone SIM card, Arlene was unlocked, and her missions began.

As Arlene, Candy travelled to Taiwan for the CIA. She maintained that she was tortured in order to test the effectiveness of the technique, and that the CIA planned to have her commit suicide once she'd served her purpose. She even wrote to her attorney to cover herself in case she disappeared suddenly or died under unusual circumstances. Sceptics claim that Candy simply suffered from trauma-induced schizophrenia, or false memory syndrome, and that her husband, the radio host Long John Nebel, who hypnotized Candy and recorded the taped interviews, was a known practical joker.

But her story attracted attention after the public disclosure of MKULTRA in 1977, and, given the evidence that has been released concerning the programme, it really doesn't seem all that far fetched. After all, if you can turn an ordinary marine lieutenant in to a Communist, why not go one step further and create a beautiful, glamorous, brain-washed super-spy.

Model and programmed secret agent Candy Jones in 1942.

PART FOUR:
SERIAL KILLERS & WAR CRIMES

The Most Prolific Killer in History

Harold Shipman (1946 – 2004), preyed on elderly female patients – the very people he had promised to protect and heal. A doctor who nurtured an unhealthy fascination with the effects of diamorphine – he saw himself as a god-like figure whose medical knowledge afforded him power over life and death.

On 24 June 1998 Dr Harold Shipman signed the death certificate of 81-year-old Kathleen Grundy, one of the patients of his medical practice in Hyde, Cheshire. It was something he had done on numerous occasions in the past, an unpleasant but necessary part of any doctor's job. Her death came as an enormous shock to her family. Considering her age, she had appeared to be in perfect health. Dr Shipman visited her a few hours before she died and was the last person to see her alive. The visit had nothing to do with any health worries. He asked if he could collect a blood sample from the old lady for a study he said he was conducting into the health of his elderly patients. At the time there were no suspicious circumstances to her death. She was an elderly woman and, on her death certificate, Shipman gave the cause of death simply as old age.

A few weeks later, Kathleen Grundy's will was read out to, among others, Angela Woodruff, her daughter. The main benefi-ciary, receiving all of Mrs Grundy's money, and her house, was Harold Shipman. Wood-ruff's suspicions were aroused. Her mother had never mentioned leaving anything at all to Shipman in her will. She contacted the police and an investigation began. The typewriter used to draw up the will was found to be one owned by Shipman.

Forged Wills and Fake Cremations

It was not the first time questions about Shipman's conduct had been asked. A few months previously Dr Susan Booth, who worked for a different medical practice in Hyde, had raised concerns about him. Over the past few years he had asked her to coun-tersign an unusually high number of crema-tion certificates, which are required under British law before a cremation can take place. A funeral director told her that, when he went to pick up the bodies of elderly women after Shipman signed a death certificate, he found many of them were fully clothed and either sitting in a chair or lying on a sofa. In his experience, people died in a wide variety of situations. The similarities between the circumstances of the deaths of so many of Shipman's elderly female patients struck him as strange. Dr Booth, already worried about the high death rate of his patients, was now convinced something was wrong. She was not sure if it was due to incompetence or if there was something more sinister going on, so she decided to take her concerns to the local coroner. A police investigation began, but the officers assigned to the case found no evidence of malpractice. They couldn't have looked very hard.

The discovery of the forged will caused the police to begin a much more serious investigation. They obtained a court order to exhume Kathleen Grundy's body, and an examination revealed traces of diamorphine in her body. There was no medical reason why Mrs Grundy should have been taking diamorphine – an extremely powerful painkiller analogous with heroin. It would normally only be prescribed in very serious cases, such as during the treatment of terminally ill cancer patients.

Overdosing on Diamorphine

Shipman was arrested on 7 September 1998. It became apparent very quickly that he had killed many of the patients in his care with overdoses of diamorphine, then forged their medical records to make it look like they had been sufficiently ill to warrant the treatment. It was not a question of whether or not Shipman murdered some of his patients. The question was, how many of them did he kill? Some undoubtedly died of natural causes and some of the bodies were cremated, making it impossible to say for sure how they died. Exhuming all the bodies of those who were buried was a daunting prospect. Potentially there were hundreds of cases.

On 5 October 1999 Shipman was put on trial for the murder of 15 elderly women who

Harold Shipman, nicknamed 'Dr Death' after his horrific killing spree came to light, was convicted of murdering 15 of his patients and sentenced to life in prison. He was found hanging dead in his cell on 13 January 2004, the day before his 58th birthday.

died between 1995 and 1998 while in his care, including Kathleen Grundy. The police described the case of Kathleen Grundy as a sample, which would be tried first so the jury didn't become overburdened by the extent of the evidence being presented to them. It was thought there could be in excess of 200 more cases to follow. The trial lasted for four months and, in January 2000, the jury found Shipman guilty of all 15 murders. He was sentenced to 15 consecutive life sentences, with a recommendation from the judge that he never be released.

The Ultimate Terminator

Four years later, on 13 January 2004, with the investigation into his other victims continuing, Shipman committed suicide in his prison cell by hanging himself from the bars of his window using torn up bed sheets. At no stage between his arrest and suicide did Shipman offer any form of explanation for the murders. David Blunkett, the Home Secretary at the time, famously said that, on hearing the news of the suicide when he woke in the morning, he wondered if it was too early in the day to open a bottle of champagne.

It is impossible to say for sure what motivated Shipman to kill so many of his patients. Many commentators point to the relationship he had with his mother – one of the obvious starting points in the consideration of any serial killer. Vera Shipman was probably the only person in Shipman's life to unconditionally believe in his abilities, with the possible exception of his wife who continues to protest his innocence. When Shipman was growing up, his mother constantly told him he was a bit better than everybody else, instilling a sense of superiority in him that he would carry throughout his life. Many of the people who worked with Shipman during his medical career found him aloof and arrogant, although he is said to have had an extremely good bedside manner, particularly with elderly women.

A Fatal Combination

In 1963, when Shipman was 17, his mother developed lung cancer. He spent many hours sitting by her bedside. As the disease progressed, her doctor gradually increased the dosage of diamorphine he prescribed to her for the pain, up until she died in June of that year. It is always easy to speculate in hindsight, but, perhaps it was during this traumatic period that Shipman developed a fascination with drugs – how they can be used to control pain and, in certain circumstances, to control life and death. It was also when he decided to pursue a career in the medical profession. Shipman's sense of superiority, together with a wish to control other people's lives and, ultimately, their deaths, would prove to be a fatal combination. After he became a general practitioner, he over-prescribed diamorphine to those patients who actually needed it, hoarded the excess and, when he decided one of his patients had lived for long enough, killed them with an overdose.

It is not known for certain exactly when he first began to kill his patients. He could have been using overdoses of prescription drugs to kill people throughout his career, beginning in Pontefract General Infirmary, West

Yorkshire, in 1970. Shipman stayed there for 4 years, before, in 1974, becoming a general practitioner in Todmorden, a small town in the Calder Valley of West Yorkshire. While there he quickly gained a reputation as a hard worker. The other doctors in the practice particularly appreciated his willingness to take on elderly female patients, who, they often found, were the most difficult and unrewarding patients to deal with. Shipman was very popular with his elderly patients. He was not as impatient as some of the other doctors, taking the time to sit by their beds and talk to them.

A Taste of his Own Medicine

In July 1975 Shipman was found to be writing prescriptions for large quantities of pethidine, a powerful painkiller sometimes known by the brand name Demerol. He was injecting it himself and had become addicted. After voluntarily entering a drug rehabilitation clinic, and pleading guilty to charges of forging prescriptions, he was given a second chance by the British Medical Council. Rather than strike his name from the register of doctors, which would have meant he would never have been able to practice as a doctor again, he was fined and warned about his future conduct.

After coming out of the clinic, Shipman worked for 18 months for the local heath authority in Durham. In 1977 he went back into general practice at the Donneybrook Medical Centre in Hyde. Although not particularly popular with other members of staff, who found him arrogant and bad-tempered, once again he took on many of the elderly female patients. In 1993 he split from the Donneybrook practice, setting himself up in a clinic on his own a few hundred yards away. Many of his elderly female patients followed him to the new practice. It would later become apparent that, during the five years he worked on his own up until his arrest in 1998, the death rate among his elderly patients rapidly increased.

Getting Away with Murder

The Shipman Inquiry, headed by Dame Janet Smith, was set up after Shipman was convicted in 2000. It led to the publication of a number of reports into how Shipman had got away with murder for so long. The last of the reports, which came out in 2005, made recommendations to the British Medical Council concerning changes to the procedures followed by doctors in the event of the death of a patient. It was an attempt to make sure that no doctor would ever have the opportunity to repeat the crimes committed by Shipman.

The report also contained an estimate of the likely number of Shipman's victims, based on how many patients could be expected to die of natural causes during a doctor's career. Over his whole career, 459 patients died while in Shipman's care, leading to the report estimating he murdered at least 250 people. It will probably never be known for certain how many people Shipman murdered in total, but, if the figure of 250 is accepted, then Harold Shipman, a doctor entrusted with the care of the sick, becomes the most prolific serial killer in history.

Murder and Mutilation in the name of Medicine

Herman Mudgett aka Dr HH Holmes (1861 – 1896). To enter the mind of Herman Mudgett would have been a fascinating, if not harrowing experience. How could a man appear to be an upstanding member of the community, a devoted husband and successful local businessman while concealing such monstrous acts of murder and mutilation?

The details of Herman Webster Mudgett's life are so bizarre and appalling as to be almost beyond belief. On the surface he was a successful doctor, pharmacist and businessman, but this facade concealed gruesome and macabre secrets which only came to light after he was investigated for insurance fraud. Eventually he confessed to the murder of 27 people, but, as he was a compulsive liar, it is impossible to know how much truth there was in his confession. Some estimates have put the true number at over 200.

Dr HH Holmes Emerges from the Shadows

Mudgett was born into an affluent family in Gilmanton, New Hampshire, in 1861 He studied medicine at the University of Michigan in Ann Arbor and, while still a student, began his criminal career with body-snatching. He would take out a life insurance policy under a bogus name, steal a corpse from a cemetery, disfigure it with acid so it could not be recognized and pass it off as the person insured. After almost being caught, he left Ann Arbor and assumed the name of Dr HH Holmes. He was involved in a variety of scams: more insurance fraud, forgery, embezzlement and, more than likely, murder, before he resurfaced as a pharmacist in Chicago. The owner of the pharmacy, an elderly widow, disappeared not long after he arrived, and he was left to run the business himself.

A string of successful ventures: selling a cure for alcoholism, inventing a machine to make natural gas, together with more fraud, made him a wealthy man. At about this time he married Myrta Belknap, no doubt neglecting to mention he had married a woman when he was 18 and never divorced her. He built a large hotel with more than 100 rooms and opened it in time for the Chicago World's Fair in 1893. It had been built by a succession of different builders, so nobody except Holmes, as he was now called, knew about all the special features he had included in the building. There were secret passages, rooms with no windows and chutes leading down to the basement. Doors opened onto brick walls, staircases led to nowhere. Business was brisk with visitors coming to Chicago for the fair or to look for work and staying in this strange hotel, which became known as The Castle.

Counting Them In and Counting Them Out

How many of the residents checked in but never checked out has never been accurately

TOP Herman Mudgett aka Dr HH Holmes.

BELOW HH Holmes' Murder Castle on 63rd Street, Chicago, Illinois. With over 100 rooms the 'Castle' was a maze of secret rooms filled with gruesome torture equipment. Holmes asphyxiated his 'patients' in a lethal gas chamber with a trapdoor. Once he had finished with a corpse it slid down a chute into a vat of acid in the basement.

established. Young, single female guests would stay a few nights and never be seen again. Waitresses and chambermaids in the hotel would leave their jobs suddenly without saying goodbye. Meanwhile, Holmes was running a side business selling skeletons to doctors and university laboratories. He even had an assistant to help him strip the flesh off the bodies and prepare the skeletons for sale. The bodies were, he claimed, those of patients of his who had died. No one appeared to have been suspicious of his activities. He was charming and gracious, particularly with women. He could talk his way out of settling the bills for the chemicals he used in his business and explain away the chemical smells coming out of the basement when guests in the hotel complained.

Eventually Holmes's many creditors demanded repayment of a large sum of money that he owed them. He duped a woman from Texas into transferring the title of the land she owned in Fort Worth to him and then killed both her and her sister, entered into another bigamous marriage and set fire to the hotel for the insurance, before skipping town with his new wife and Benjamin Pitezel, a man who worked for him in the hotel. Detectives from the Pinkerton Agency investigating various insurance claims trailed him to Fort Worth, St. Louis and on to Philadelphia, where they finally caught up with him. He had been involved in yet another insurance scam with Benjamin Pitezel, who then disappeared. Holmes produced a body and tried to claim the insurance. It was one step too far. He was challenged and came up with a story saying the body wasn't Pitezel, suggesting he was committing insurance fraud not murder. One of Pitezel's

daughters identified the body and Holmes was arrested for his murder. The Pinkerton agents continued to follow the trail Holmes had left, finding the bodies of Pitezel's wife and other members of his family in a house in Toronto, and the Chicago police began to investigate the hotel.

Entering the Gas Chamber

Inside they found airtight rooms with gas pipes running into them, and doors that could only be opened from the outside. They concluded that Holmes had killed people by asphyxiating them with gas. He then transferred the bodies down to the basement, where he had installed a dissecting table and acid baths. After selling the skeletons of his victims, he disposed of the remains of the bodies by cremating them and throwing them into lime pits. The police found various body parts in the basement, along with women's shoes and clothes.

Holmes was put on trial for Pitezel's murder, even though details of at least some of his other crimes were beginning to emerge. He was found guilty and sentenced to death. While awaiting execution he wrote a long and rambling confession to 27 murders and said he thought he was possessed by the devil. The *New York Times* reported his execution, saying he had been calm and collected immediately beforehand, asking the hangman to take his time and to do the job properly. The hangman doesn't appear to have listened and Holmes died slowly. After the trap in the gallows opened and he dropped to the end of the rope, it was 15 minutes before he was pronounced dead.

Human Experiments

Sigmund Rascher (1909 – 1945). Nazi medical men flagrantly defied the code of ethics throughout World War II and Sigmund Rascher, the cold-hearted Doctor of Dachau was no exception. Possessing a heart as ice cold as the baths into which he submerged his subjects, he performed chilling experiments on the camp's inmates, forcing them to undergo torturous tests for the dubious benefit of German soldiers.

Stormtrooper Sigmund

On 12 February 1909 the city of Munich in southern Germany witnessed the birth of one Sigmund Rascher. With its motto of *Weltstadt mit Herz* (A cosmopolitan city with a heart) the Bavarian capital seems an unfitting birthplace for such a notorious figure, who would become known for heartless acts upon his fellow man. Born the third child of a physician, it was somewhat inevitable the boy would follow in his father's footsteps, entering the world of medicine through his studies at his home university.

After a short stint in Basel, Switzerland, where he worked with his father, Sigmund returned to Munich to complete his education, receiving his doctorate in 1936. In May of this year he made his political position known by donning the brown shirt of the *Sturmabteilung*, the Nazi party's stormtroopers. This did not impede his continued medical researches.

Keen to promote the anthroposophic theories of his spiritual guru, Ehrenfried Pfeiffer, he worked on a copper chloride test to detect pregnancy and cancer cells. The astounding results aroused the suspicion of the authorities and an independent analysis was ordered. If his findings were discovered to be fraudulent, it would spell the end of his burgeoning career. Luckily for Doctor Rascher, he made a powerful friend in the Nazi hierarchy soon after that prevented any comebacks.

Himmler and Her

Thanks to his intense medical studies, Sigmund was already a well-connected physician, counting professors and scientists from all over Germany as close friends. But in 1939 he formed an alliance with one man that would far outweigh all other connections. His name was Heinrich Himmler, the infamous Reichsführer of the SS.

At this time Rascher was dating former

cabaret singer, Karolina 'Nini' Diehl. Older than him by 15 years, the one-time chanteuse had been mistress to Himmler, over whom she still held considerable influence. The police chief often indulged her many whims and so when his erstwhile lover asked for help with her new partner's career he happily obliged. Rascher was elected to the *Ahnenerbe* – an elite Nazi think tank – and given the honorary rank of *SS-Untersturmführer*.

Granted direct access to Himmler, Nini's sweetheart was able to bend the fellow Munich man's ear when it came to his scientific studies. In Spring 1939, Rascher asked for assistance with his private cancer research. Up to this point he had been trialling plant extracts as a treatment on lab rats but the doctor believed it would be more efficacious if they could switch to human subjects. His superiors had deemed the move as inhumane but Himmler, who was already bankrolling the studies, eventually sided with the researcher.

An All Time Low

Sigmund's search for a cancer cure was put on hold when war broke out in September 1939. He joined the *Luftwaffe*, the German air force, as a junior doctor working in the medical services. It was in this role that he encountered the high-altitude tests being carried out in the aviation medicine department. To improve the survival rate of German pilots forced to eject thousands of feet up in the air, monkeys were submitted to depressurization procedures to ascertain the effects of such an experience.

In May 1941, Rascher wrote to his esteemed contact suggesting they could improve these experiments by using human subjects. Lamenting the lack of volunteers for such a dangerous exercise, the doctor petitioned for two or three feeble-minded prisoners to be placed at his disposal. The *SS-Reichsführer* promptly acceded to the request and the logistics for such an operation were developed at a medical conference in early 1942.

Testing began in earnest in March of that year at Dachau concentration camp 10 miles north-west of Munich. Around 200 prisoners were selected at random to undergo a series of experiments inside a portable pressure chamber. This 2-metre-high cubicle was made of wood and metal and allowed the researchers to view the subject through a small window. Once locked inside, the victims would be exposed to a drop in pressure simulating a descent from up to 68,000 feet.

Four separate versions of the test were conducted reproducing a fall with and without a parachute, with and without oxygen. The human specimens would suffer extreme pain, foam at the mouth and often lose consciousness. Rendered comatose, these victims underwent dissection, their skulls split open to understand the low pressure effects on the brain. It was reported only 40 survived these experiments, the remainder were executed.

Immersion Aversion

Following the high-altitude tests, Rascher wanted to tackle another problem that Nazi fighting men faced during the war: the extreme cold. Many German pilots were forced down into the freezing North Sea during sky battles with the enemy. Also, infantry at the Eastern Front had to contend with the bitter conditions of a Russian winter. It was essential they discovered a means to combat the low temperature in both instances.

To this end, Rascher suggested a series of experiments to discover the best method for rewarming the human body. After Himmler gave his approval, tests began on 15 August 1942. They were divided into two phases. First, to establish the length of time it would take to lower the body temperature to incur death; second, how to best resuscitate the frozen.

There were also two methods of lowering the body temperature. Either the human subjects were forced to remain outdoors, strapped down on a stretcher, ice-cold winds punishing their bodies for up to 14 hours at a time. Or they were sent to the bath. This was a large 8 metre square tank of ice cold water, in which the chosen victims would sit dressed in full-flying gear complete with life jacket.

Using a stethoscope strapped to the skin to measure the heart-rate and a rectally-inserted probe to monitor the body tem-

Dachau Concentration Camp, Germany (1942). Dr Sigmund Rascher (right) conducts a hypothermia experiment on an inmate. The test subject was immersed for 3 hours in icy water.

perature, Rascher and his assistants observed the effects of the sub-zero conditions. With persistent experimentation Sigmund managed to lower the heart rate to a torpid four beats per minute, quickly learning that dropping the body temperature to 28° Celsius was sufficient to kill the subject.

Next came the rewarming. A variety of cruel methods were devised by the doctor to raise the body temperature of his human guinea pigs. Hot sun lamps – so hot that they seared the flesh – were tried. Scalding water forcibly injected into the stomach and bladder to heat from the inside out fared little better. Even Himmler, on a rare visit to view the experiments, suggested using sexual stimulation to warm the frozen victims. Gypsy women were promptly sent from Ravensbrück, stripped naked and forced to act as human blankets, often resulting in copulation.

In the end, the best results came from immersion in hot water. Over 400 experiments on around 300 human specimens had taken place to make this discovery. Autopsies on the still-living ensured around a third of the subjects perished. Rascher then presented his findings at a medical conference in Nuremberg in October 1942, inciting protests from all sections of the armed forces. This had no effect. Still backed by Himmler, the Nazi doctor was permitted to continue with his inhumane tests.

Kidnapped Kinder

Rascher persevered with his research, making further use of the Dachau prisoners. Believing he had discovered a blood-clotting preparation called Polygal, he began testing its properties on human specimens. They were given a dose of the pectin substance and then shot through the neck, had their limbs amputated or even infected with bacteria to gauge the efficiency of the drug. Soon after, he set up a company to manufacture the product, staffed by camp inmates. Many doubted the coagulative qualities of the drug, confirming it was nothing more than a salt solution. Once again, Sigmund's findings were being regarded with suspicion.

Then in May 1944 his fraudulent ways caught up with him. He and his wife had long been revered as poster parents for the procreation of the Aryan race after Nini had given birth to three blonde-haired children after the age of 48. In fact, all three had either been illegally bought or kidnapped from orphanages. Feeling personally affronted, Himmler had them both arrested. Further investigation revealed Sigmund had doctored many of his experiments' results, fabricating data to support already-held theories.

Deemed undeserving of a trial, the pair were executed on the direct orders of the irate SS chief. Nini was hanged at Ravensbrück prison camp while her husband met his end at Dachau, the home of his cold-blooded tests, on 26 April 1945.

Three days before liberation, Sigmund Rascher received a bullet to his brain from the pistol of *SS-Hauptscharführer* Theodor Bongartz in cell 73 of the camp's bunker. A month later, from beyond the grave, Rascher would exact revenge on his erstwhile patron. Following his arrest in Lüneberg, Heinrich Himmler committed suicide by swallowing a cyanide pill – a capsule developed by the demonic doctor.

Children of the Fatherland

Josef Mengele (1911 – 1979). Known notoriously as the Angel of Death, this military hero turned medical monster subjected the prisoners of Auschwitz-Birkenau to an unrestrained miscellany of obscene experiments – all in the name of science. Collecting twins in his own private zoo, the candy-carrying killer called Uncle Mengele by his unsuspecting charges injected, operated and ultimately murdered in a bid to double the birth rate of the blonde-haired and blue-eyed children who were the pride of the Nazi nation.

A Mentor for Mengele

As a child young Josef Rudolf Mengele never walked in the shadows. Born on 16 March 1911 to a distant father and a cold, unloving mother he was always striving for his voice to be heard. As he grew up in the Bavarian river city of Günzburg in southern Germany he struggled against his parents' apathy to develop an innate belief that he was destined for greatness. The ambitious Beppo, as he was nicknamed, even declared to friends that his name would end up in an encyclopedia.

Much to his father's disappointment, Josef shirked his responsibility as the eldest son to carry on in the family business – the manufacture of farm machinery – and in 1930 enrolled at the University of Munich. Five years later he gained his PhD in Anthropol-ogy with a thesis on the racial differences of the lower jaw. Still in his teens, Josef had a clear passion for eugenics, a science focused on improving hereditary traits of the human race. And he further pursued this area of study by taking an assistant research post at the Institute for Hereditary Biology and Racial Hygiene in Frankfurt.

Here he formed a close bond with Dr Otmar von Verschuer, a leading light in the field of genetics. Enraptured by the professor's work seeking out the keys to unlock hereditary secrets Mengele followed in his mentor's footsteps, studying human deformities and abnormalities. Together they hoped to find a way to iron out genetic defects and undesirable characteristics to ultimately purify the German bloodline.

Fighting for the Fatherland

The young researcher saw von Verschuer as the father figure he had always longed for and was soon beguiled by his political leanings as well as his scientific work. His fascist mentor saw his pupil join the Nazi party while working in his employ. A year later, Mengele had become a member of the SS and, three months after obtaining his medical degree, spent a period of time combat training in the Tyrolean mountains. The Bavarian doctor was not all brains, it seemed, but brawn too.

When war broke in September 1939, Mengele was itching to fight for his country. But the eager SS-man was forced to wait. A past kidney complaint delayed his journey to the front until the summer of 1940 when he joined the medical corps of the *Waffen-SS*. A year later, Hitler declared war on Stalin's Russia and Lieutenant Mengele was sent to the Ukrainian Front where he excelled as a soldier, receiving the Iron Cross Second Class for bravery.

In January 1942, whilst serving with the *SS Wiking* Division behind enemy lines, he dragged two German soldiers from a burning tank. This act of heroism earned him three more honours including an Iron Cross First Class. Injured during this engagement, Josef was forced out of the war, declared medically unfit for combat, and was shipped back to Germany spelling the end of his fighting days.

At this time, he recommenced his partnership with Verschuer who was now at the Kaiser Wilhelm Institute for Anthropology, Human Genetics and Eugenics in Berlin. From fields of snow dodging Soviet snipers, Mengele was back in the field of science and was about to receive an assignment that would ensure the inclusion of his name into all encyclopaedia.

Arrival at Auschwitz

In late May 1943, Doctor Josef Mengele arrived in south-west Poland at one of the Nazis' most notorious creations: Auschwitz concentration camp. Given the position of physician to the gypsy camp at Birkenau (the extermination annex less than two miles away) the newcomer was placed in charge of the well-being of the Romany people who were being transferred in their thousands from ghettos right across Europe.

However, this medical officer had more on his mind than the health of the prisoners. His mission was to continue his mentor's work in genetics, to devise new ways of eliminating unwanted gene strands in the search for racial purity. Almost immediately, Mengele got to work on a gangrenous disease called noma that had spread throughout the gypsy camp.

He began experimenting on the infected, executing victims in order to perform autopsies on their diseased cadavers, hunting for potential hereditary links. Dissected organs and decapitated heads were preserved in jars and sent to outside institutions for analysis. Less than eight months after his arrival, the entire gypsy camp was liquidated. While his charges were mercilessly received en masse into the gas chambers, Mengele received a promotion to chief camp physician at Birkenau.

Mengele's Children

The camp physician became an ever-present at the Birkenau train station; the end of the line for endless new arrivals. Dressed in a pressed green suit replete with medals, his white-gloved hands holding his trusted riding crop, Mengele cut a genteel figure before the crowds of dishevelled passengers. Whilst whistling a favourite aria, he determined the fate of the fallen. With a cursory glance and a simple call of *links oder rechts* (left or right) he assigned the ushered and unwilling to either a life of hard labour or a short trip to the smoking crematoria.

The bulk of the prisoners were of no interest to the death-giving doctor. He was on the look out for ideal test subjects, specifically twins. Produced from the same gene pool made them the perfect specimens for his heredity experiments, enabling him to contrast and compare reactions to a va-

riety of stimuli. Furthermore, Mengele was commissioned to discover the genetic cause for the twins in order to increase the Aryan population. Find the method, double the birthrate.

Following on from his mentor's own research, Mengele conducted the most invasive and excessive experiments upon these twins, many of whom were young children. Unhindered by a need or desire to adhere to any code of ethics, he snatched them from their parents and held them in a special barracks nicknamed 'the Zoo'. Initially, they were afforded special treatment. Uncle Mengele, as many of the minors called him, would hand out sweets with a gentle smile. But this kindness was purely a façade.

To begin with, the twins underwent a preliminary examination. Every inch of their bodies was measured in the minutest detail before they were tattooed and assigned to

Dr Josef Mengele (left) at Auschwitz Concentration Camp in 1943. Next to him is Rudolf Hoss, Commandant of Auschwitz. Second from right is Josef Kramer, Commandant of Belsen. Right, an unidentified German officer.

their new home. Here each child awaited an unknown fate.

Seemingly selected on a whim, Mengele's Children endured despicable act after despicable act. Spinal taps, X-rays and chemical smears left them in agony. Transfusions from twin to twin despite dissimilar blood types caused terrible headaches and high fevers. Limb and organ amputation and exchange between siblings were performed without anaesthetic. His experiments knew no bounds, even sewing together a pair of female twins, artificially conjoining them via their hands which soon became gangrenous due to infection.

When Mengele had finished testing, he would deliver the *coup de grâce*: a shot of chloroform or phenol to the heart, killing them instantly. He would then dissect the bodies to learn of the internal effects of his abhorrent experiments. It is thought of around 1500 pairs of twins he persecuted only 100 pairs survived the torment.

After Auschwitz

While the twins were his most treasured specimens, Mengele did have a passion for other human anomalies. He also studied those affected by dwarfism. The Ovitzes, a family that included seven dwarves, were among his prized possessions. Separated from the other inmates, they were subjected to procedures typical of this Birkenau butcher. Bone marrow was extracted, teeth pulled and gynecological tests carried out to understand the physiological reason behind their perceived deformity.

Mengele was also fascinated by those with heterochromia, a condition whereby a person's irises differed in colour. He injected various chemicals including methylene blue in an attempt to alter eye colour and even harvested a collection which he sent to a specialist in Berlin. Pregnant women also caught his eye and were operated on with impunity.

Finally, after nearly three years of unchecked experimentation, Mengele's reign of terror came to an end. Fleeing before approaching Allied troops, the death camp doctor packed up his medical records and headed west in January 1945. After some time on the run, living under the alias Fritz Hollmann, he was caught but his American captors were unable to positively identify him and released him. For the next four years he worked as a farmhand in Bavaria until eventually escaping war-torn Europe for sympathetic South America.

The Final Hiding Place

For the next 30 years Mengele remained hidden along with fellow Nazi émigrés in Argentina, Paraguay and lastly Brazil. It was here in the beach resort town of Bertioga that he met his end. While swimming in the ocean, he suffered a stroke and drowned on 7 February 1979. Buried under the name Wolfgang Gerhard, it took a further six years for West German investigators to locate him. The grave was exhumed and forensic analysis confirmed the remains belonged to Josef Mengele. Many still suspected the devil doctor had tricked the world into believing he had died until 1992 when a DNA test proved unequivocally that the Angel of Death was dead. He would live on only as an encyclopedia entry.

Unit 731

Shiro Ishii (1892 – 1959). During World War II a top-secret germ warfare centre was set up outside the city of Harbin, China. The cover story was in place: this huge compound was carrying out essential work in the field of water purification. In reality, Unit 731 was performing horrifying experiments on prisoners of war, all at the hands of the corrupt sadist, General Shiro Ishii.

Shiro Ishii, born 25 June 1892, was a bright but odd child. He was a loner, and his selfish nature drove away any friends he made. His extreme intelligence led him to feel superior to his peers, and this isolated him from many personal relationships; but his confidence and self-determination worked in his favour later in life.

As an adult he worked at the 1st Army Hospital and Army Medical School in Tokyo where he impressed his superiors with his aptitude for medicine and his obsequious nature. Later, he studied at the Kyoto Imperial University, graduating as a doctor in 1920. In 1925 the Geneva Convention banned the use of bacteriological and chemical weapons in war; however, this did not prohibit countries from researching or producing such weapons, and Ishii began to push for the establishment of a Japanese bacteriological weapons programme.

He was spurred on by the belief that if such weapons were banned, they must be very effective. In 1928 he began a two-year tour of Europe and the United States, during which time he researched the effects of biological and chemical warfare post-World War I. When he returned to Japan he was promoted to the rank of major and began work at the Tokyo Army Medical School.

He continued to push for the development of a bacteriological weapons programme, and in 1936 was set up in a facility, the pleasant-sounding Anti-Epidemic Water Supply and Purification Bureau. In the future, this facility was to become world-famous as Unit 731, a compound synonymous with stomach-churning atrocities.

The Asian Auschwitz

The site at Pingfan near Harbin was an enormous area, covering 6 sq km (2.3 sq miles), housing 150 buildings and employing 3,000 personnel. Here, Ishii was expected to research and develop his bacteriological weapons programme, and invent a dispersal mode for a devastating killer chemical agent.

Ishii was a sadist at heart, and his cruel methods knew no bounds. His keen interest in bacterium, and its deployment in warfare led him to embark on a series of evil human experiments. American, Chinese, Soviet, Korean and British prisoners of war, as well as the elderly, children and women, were used in his experiments, each forced to suffer repulsive procedures being tested like helpless animals.

Some were vivisected unanaesthetized, screaming in agony, as Ishii performed limb amputations to research the effects of blood loss. Ishii would often reattach an amputated limb to a different site on the prisoner's body.

Internal organs would be surgically removed and replaced in another area, while the live prisoner suffered incomprehensible agony. It was common for prisoners to be routinely shot, to allow Ishii's surgeons to practise bullet removal.

Other poor wretches were tied to stakes and bombarded repeatedly with shrapnel laced with gangrene so Ishii could record the rate of infection.

Pushing the Limits of the Human Body

Prisoners were locked in pressure chambers while researchers timed how long it would take for their eyes to pop out of their sockets. Some were submerged in freezing cold water, or locked in tanks and exposed to sub-zero temperatures. Researchers monitored how long it took for frostbite to develop or hypothermia to set in.

Prisoners were also hung upside down or placed in large centrifuges and spun rapidly, to find out how long it took them to die. Horse urine was injected into the kidney and animal blood injected into the bloodstream, to study the effects.

Ishii was also keen to psychologically test his prisoners, and allegedly many female

Shiro Ishii, microbiologist and commander of Unit 731, the biological warfare unit and horrific slaughterhouse of the Imperial Japanese Army.

prisoners were impregnated by rape and then forced to undergo savage abortions. Venereal diseases such as syphilis and gonorrhea were also transmitted via rape or injection, and the side-effects studied.

Other cruel experiments involved depriving prisoners of food and water, and recording how long the body can survive without nourishment. Sleep deprivation and being burnt with acids were also standard procedures.

Home-Brewed Bubonic Plague

Ishii, being a trained microbiologist, was fascinated with bacteria and viruses, and relished this opportunity to work with diseases. Prisoners would be deliberately infected with home-brewed anthrax, smallpox, botulism, cholera, bubonic plague and other deadly pathogens, and the effects on the mind and body would be ardently observed.

Healthy prisoners would be quarantined with sick ones, allowing the researchers to chart the speed in which disease spreads. Once a prisoner died, any organs would be harvested and the corpse tossed into the incinerator like garbage.

Anthrax-Infected Flea-Bombs

Prisoners would be forcibly infested with disease-infected fleas and be used as breeding grounds for the tiny insects. The goal was to investigate the viability of using fleas in germ warfare, something which Ishii soon did. Pingfan had 4,500 flea breeding machines which were capable of producing up to 100 million fleas every few days. Ishii would infect these fleas with lethal pathogens such as cholera and anthrax, and encase them in bombs. The army would drop these over villages, causing outbreaks of disease. It is estimated that approximately 400,000 Chinese civilians died as a result of these flea-bombs.

Taking the Grisly Truth to the Grave

At the end of the World War II, Japanese troops were commanded to blow up Unit 731 to cover up what had taken place. Ishii ordered the murder of the remaining 150 prisoners at the camp, telling them they would 'take their secret to the grave'.

Personnel at Unit 731 were threatened effectively enough to keep their mouths shut, and Ishii fled to Japan.

In 1946, the United States occupied Japan and began negotiating with known war criminals. They tracked down Ishii and proposed a deal. He would be given full immunity from prosecution for his crimes against humanity, in exchange for the data, from his grisly experiments.

This was an ethically questionable decision for the US government. They did not approve of Ishii's methods, but felt his knowledge would be invaluable, and safer in their hands.

Ishii accepted the deal, and became a free man, living a good life until his death from throat cancer in 1959. It has been estimated that 200,000 prisoners died in the gruesome experiments at Unit 731, all in the name of science, and of furthering Japan's arsenal of biological and chemical weapons.

The Atomic Bomb

Enrico Fermi (1901 – 1954). Along with Robert Oppenheimer, Enrico Fermi is often referred to as the father of the atomic bomb. His work on induced radioactivity was so significant that in 1938 he was awarded the Nobel Prize for Physics. But Fermi's research also had horrifically evil capabilities which would go on to kill tens of thousands of people and change the world forever.

Enrico Fermi was born in Rome in 1901 and became an expert in mathematics, physics and engineering. He was first and foremost a professor of physics, later of atomic physics. Suffering persecution in Italy because of his Jewish wife and his expressed anti-fascist views Fermi took the opportunity of the Nobel Prize award ceremony in Stockholm to escape from Italy. He just didn't go back, but went to New York instead.

It was at Columbia University in New York that Fermi started working with Leo Szilard and a graduate student called Walter Zinn. Together they started experimenting with nuclear fission, on the assumption that the neutrons released during fission would begin a chain reaction – and generate nuclear energy.

The research Fermi and his co-workers were engaged in clearly had the possibility of being developed for military use. In March 1939, Fermi tried to discuss the progress of his research with the US Navy, though nothing came of the encounter. Then Szilard explained to Albert Einstein what they were doing. Einstein was an extremely influential lobbyist.

Einstein steps in

It was Einstein who alerted President Roosevelt to the military implications of the project, and suddenly lavish funding and a team of workers were made available in the cause of World War II. Fermi's project was monitored by a second team of scientists from Princeton University. The Princeton team gave its approval and the joint team, based in Chicago, got to work on the nuclear energy project.

It was in December 1942 that the first experiment in controlled nuclear fission was carried out. It took place in a squash court at Stagg Field at the University of Chicago, and was a great success. In August 1944, the project was moved to Los Alamos in New Mexico, where a new tailor-made laboratory had been built under the leadership of Robert Oppenheimer.

Fermi became head of the Los Alamos

physics department. The new lab's specific goal was the construction of a working nuclear bomb that would win the war.

Hiroshima and Nagasaki take the hit

A further year's work and US$2 billion dollars later the first atom bomb was detonated early in the morning of 16 July 1945 at the Alamogordo airbase, 120 miles south-east of Albuquerque. The bomb test was an overwhelming success, and the team set about building more atomic bombs for deployment in the closing stages of World War II.

The Americans had the Japanese in retreat, but there was a danger that the Japanese would kill all their surviving prisoners, and refuse to surrender. The Americans made the historic and controversial decision to shock the Japanese into surrender by dropping one of the new weapons on a city on the Japanese mainland.

Just three weeks after the pilot test at Alamogordo, an atomic bomb was dropped on Hiroshima on 6 August. The 10-feet long bomb was delivered by the US Army Air Force B-29 bomber *Enola Gay*, piloted by Paul Tibbets.

The bomb, nicknamed Little Boy, was released at an altitude of 5,000 feet, and it exploded in the air about 600 ft above the city. The energy released during the explosion was enough to flatten completely 4 square miles of the city of Hiroshima. A total of 100,000 people in Hiroshima were killed outright. Of the survivors, perhaps another 100,000 were so horribly burnt, maimed or affected by radiation sickness that they died soon after.

The Japanese government did not respond to what was intended to be a *coup de grâce*. There was no formal surrender. The American government decided that it had no alternative but to give a second Japanese city the same treatment. The second bomb, codenamed Fat Man, was dropped on Nagasaki on 9 August. This time the atomic bomb had the desired effect, and on 10 August the Japanese government asked for a truce; the negotiation of surrender terms had begun.

The Fear of God and World War III

The invention of the atomic bomb had an immediate impact on world history. It spectacularly, if cruelly, terminated the World War II by killing tens of thousands of Japanese civilians. It also gave a terrible warning of what a World War III would be like: far worse than any war in history.

The appalling photographic images of the annihilated city of Hiroshima and its afflicted people haunted the post-war generation and hung like the sword of Damocles over the Cold War. The nuclear weapons that were stockpiled during the next few decades were even more powerful and frightening than the Hiroshima bomb.

The knowledge of the damage they could do was enough to create a global stalemate; fear of the atomic bomb became a major political force, the nuclear deterrent. As the Cold War developed, it was the memory of what the atomic bomb did to Hiroshima that put the fear of God into the superpowers and their allies. This single invention of Enrico Fermi's created half a century of global fear.

An atomic bomb test, French Polynesia. The USA conducted 216 atmospheric nuclear tests between 1946 and 1962.

Pristine paradise islands in the Pacific were totally destroyed, at times leaving a mile-wide crater on the ocean floor. Thousands of Polynesians were exiled and left traumatized. What is left of Bikini Atoll is uninhabitable to this day. Between 1946 and 1958, 23 nuclear devices were detonated at Bikini Atoll, beginning with the Operation Crossroads series in July 1946. The people of Bikini have yet to resettle in their homeland, the island is populated by Bikini Project Department construction workers and some US Department of Energy staff. There is, however, a large population of Bikinians living elsewhere in the Marshall Islands and overseas who hope to have the ability to return to their homeland someday.

106 of the atmospheric tests were carried out a mere 63 miles from Las Vegas, Nevada. The radioactive fall-out drifted eastward extending over the breadth of the nation. The countless cancer victims that these tests caused had to wait until the 1990s for a formal apology and monetary compensation from the government. By that time very many had already died.

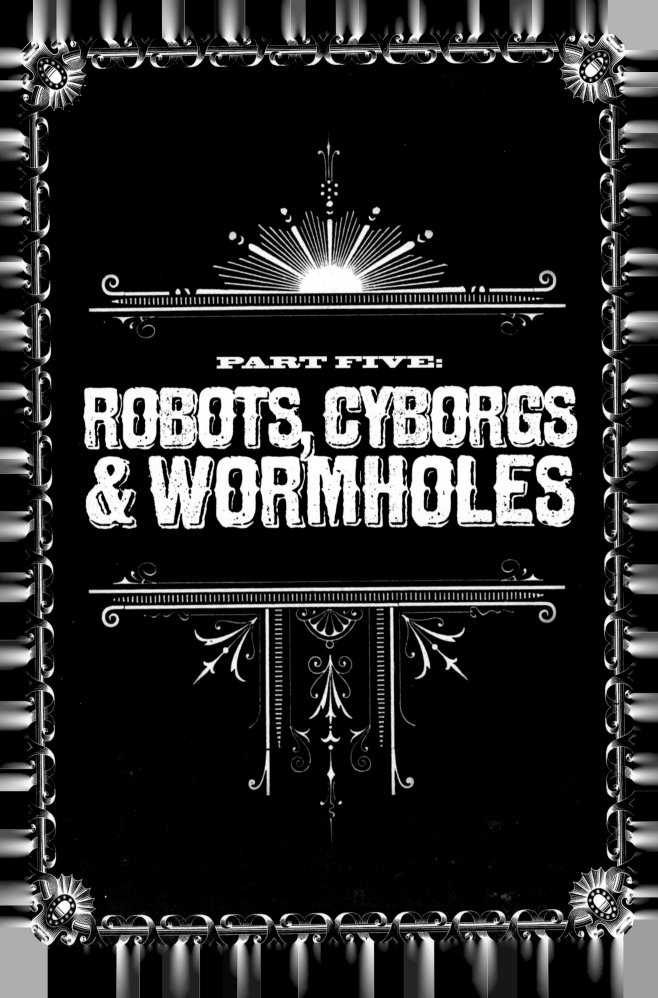

PART FIVE:

ROBOTS, CYBORGS & WORMHOLES

ALIAS CAPTAIN CYBORG

Kevin Warwick (1954 –). In 1998, a British professor of Cybernetics at the University of Reading, UK, declared he had set out a rather extraordinary mission for himself. This mission, dubbed Project Cyborg, would see Kevin Warwick living at the frontier of robotic technology, become part-man, part-machine. This is not science fiction, this is science fact.

Phase One:
The Captain Logs On

Kevin Warwick's research work in the fields of artificial intelligence, brain machine interfaces, microchips, robotics and bio-medical engineering led him to develop the idea of fusing man and machine. His aim was to test the limits of the human body, and investigate the potential of future human utilization of robotic technology.

Stage one took place on 24 May 1998 and involved surgeons implanting a capsule containing several micro-processors into Warwick's arm. Microchips had never been surgically embedded in anyone before, and this mission earned him the alias 'Captain Cyborg'.

The procedure was a success, and Warwick spent the next nine days observing the results. He found that everywhere he went the implant was communicating with computer signals, and in the Cybernetics department at the university, machines were tracking his movements. He also discovered that lights would turn on as he entered a

room and, bizarrely, his computer would greet him as he sat down.

Warwick was amazed at the way technology responded to the microchip and how well his body played host to this chip. He began to plan phase two, which was launched four years later.

Phase Two:
You Need Hands

On 14 March 2002, Warwick received a second implant, directly into his nervous system. The microchip was fired into his left wrist, and had 100 electrodes attached to it, measuring 1.5 mm long. Warwick's arm had wires running up it from the insertion point, connected to what was known as the 'connector pad', fitted with a radio transmitter. The connector pad allowed his neural signals to be transmitted via radio waves to the computer and other pieces of technology.

One goal was to prove that this idea would allow users to remotely control or operate certain electrical items. As Warwick demonstrated, he was able to operate a robotic hand simply by using his mind, allowing the hand to mirror the movements of his own, wirelessly. In another instance, Warwick conducted an experiment whereby he travelled to New York and operated a robotic hand in the UK, live over the internet, using his neural signals.

The point Warwick was making with this, was that an amputee could, in theory, control a robotic hand and it would not necessarily have to be attached at the site of the amputation; it could be anywhere and still be controllable.

The Ultimate Mr and Mrs

Some husbands and wives claim they are already telepathic and know what each other is thinking. However, Kevin and his wife became the ultimate Mr and Mrs when Irene also had a microchip fitted in her arm.

If Kevin and Irene's microchips could communicate with each other, the couple would be able to record their emotional and mental responses to sensory experiences on the microchips. These responses could then be stored as feelings patterns and accessed and re-experienced whenever they wanted.

Using the internet to communicate their neural signals, they discovered that if Irene clenched her fist three times, her neural signals were transmitted and received by Kevin's nervous system. As soon as she made the movements, he would immediately feel them – electrical brain -to-brain telepathy.

This led Warwick to theorize that one day, all people will communicate in this way, with no need for speech.

Living in Cities of Cyborgs

Warwick's vision is that one day everyone has microchip implants giving humans the super-intelligence of machines, creating 'superhumans' living in cities of cyborgs.

Kevin Warwick claims that in the future the world will be dominated by robots with intelligence far superior to humans, and that it will be the robots that rule the world. Warwick is not scared of this future, but rather focuses on giving human capabilities an upgrade using technology, in preparation for the symbiotic existence that is ahead.

He sees a future where this technology can change how we communicate, augment and expand our senses and enhance our memories and experiences. In addition, he claims that electronic medicine is a remedy for the future, and that one day this technology will play a bigger part in medicine, compared to the chemical treatments we are currently used to.

If man and machine are to merge, some argue that there would be moral and ethical issues to consider. There are concerns over microchip units being fitted with GPS technology, and the wider implications that this has on invasion of privacy and our human rights.

Of course, the ultimate fear is that cyborg technology will, as many Hollywood Sci-Fi movies predict, lead to the demise of humankind and the eventual total destruction of the planet.

SCIENTIFIC FUSION ART

Guy Ben-Ary (1967 –). From his sophisticated laboratories, Guy Ben-Ary merges nature with technology. Combining cybernetics with cultured brain cells, and creating biotechnological artwork he forces us to question what it means to be alive.

Guy Ben-Ary was born in Los Angeles, USA in 1967, lived most of his life in Israel and currently works and resides in Perth, Western Australia. He is an artist whose work uses emerging medias and in particular biologically related technologies (tissue culture, tissue engineering, electro-physiology and optics). Since the year 2000, Ben-Ary has been Artist in Residence in SymbioticA – The Art & Science Collaborative Research Laboratory at the University of Western Australia.

Biokino and The Living Screen

In 2004, Ben-Ary formed the art group Biokino, with Tanja Visosevic and Bruce Murphy. The trio explored the interface between art, film theory and biotechnology, creating their most famous piece of work, *The Living Screen*, which made its debut in 2006 in Perth, Australia. This was a 'living' cinematic apparatus which utilized early cinematic technology to fuse biotechnology and art together.

There are three elements to the process: the bio-projector, the living screen and the nano movies. The bio-projector was modelled on the early motion picture device, the Kinetoscope, because of its creepy similarity to a coffin, and the fact that a movie could be viewed through it.

The bio-projector allowed individuals to view the living screen, one person at a time. The living screen consisted of projected living tissues and cells. It functioned as the area on which the nano movies were played. Throughout the duration of the nano movies the cells would react, transform, change shape and eventually die, causing the nano movies to die with them. This process forced the spectator to directly engage with a living, bloody canvas, and experience the nano movie contort and die.

This grim conclusion was devised, through the projected imagery rather than the narrative of the nano movie, to challenge the spectator to ponder the issues of life and death, and analyze the reality around them.

The Semi-Living Worry Doll

The Tissue Culture and Art Project was devised by Guy Ben-Ary, Oron Catts and Ionat Zurr. One of the tasks within this project was to produce 'worry dolls', a South-American child's totem in which the child confides all its worries, and then is placed under the child's pillow at bedtime to absorb the worries away.

The unique element of this task, however, was that the worry dolls were created from tissue culture, and grown in vitro, resulting in a 'semi-living' worry doll. The dolls were kept inside liquid-filled glass vials and 'fed' in order to keep them alive. If removed from the vial they would 'die' due to their absence of an immune system. The aim was to explore the capabilities of tissue technology, and whether complex organisms kept alive outside of the body can be grown into specific shapes or forms.

As with all of Ben-Ary's projects, the combination of science and art is designed to invite philosophical debate or simply to serve as a symbolic question mark. In this instance, the semi-living worry doll project became, to some spectators, an issue of ethics, and some questioned whether it was right to create a living organism just because we can.

The Transglobal Cyber-Rat

The project entitled Meart (multi-electrode array art) was another fusion of artistry and biotechnology. It took place in two separate laboratories; one in Australia, and the other in Atlanta, Georgia, USA.

In Atlanta, the cultured brain cells of a rat were grown in a neuro-engineering laboratory, and were linked over a network to a robotic drawing arm in Australia. Ben-Ary achieved his goal of making the brain cells of the rat dictate what the robotic arm would draw, creating a cyber-rat capable of communicating over a network. Ben-Ary used the internet to serve as the nervous system, linking the 'brain' of the rat with the 'body' of the robot, in effect creating a transglobal cyber-entity.

The result of this experiment was the successful merging of artificial and living elements, a concept which the human brains behind project Meart believe is key in the future. In the decades and centuries ahead, it is believed humans will manufacture entities; entirely new, technologically-engineered identities which are 'semi-living' – part-grown and part-constructed.

With these hybrid creations, of course, potentially comes emergent behaviour which may be different to the intention of the creator. If intelligent and sentient entities are produced, there is no guarantee that they will develop in the way scientists want them to. In addition, as proven by the robotic arm's drawings, art itself could become a human aptitude obtained artificially, raising major questions of human ability and identity.

CLONING DOLLY

Dolly (1996 – 2003). Dolly became the most famous sheep in the world, when she became the first animal to be cloned from an adult cell, not an embryonic one. Dolly's creation was celebrated around the world as the most significant scientific breakthrough of the decade.

Ian Wilmut and Dolly the cloned sheep. On 5 July 1996 in Scotland, Professor Ian Wilmut created a sheep named Dolly, the first mammal to be replicated directly from an adult cell.

In 1996, at the Roslin Institute near Edinburgh, Scotland, scientists Ian Wilmut and Keith Campbell were working on producing a new, genetically identical individual from a single parent animal. The Roslin Institute had explored cloning before. In 1995, Keith Campbell and colleague Bill Ritchie successfully produced two identical lambs, Morag and Megan, from embryonic cells. The animal husbandry of Morag and Megan heralded the next development in the field of cloning.

The scientists were then determined to try the process again but with two different aims. Firstly, to produce a mammal from an adult somatic cell, not an embryonic one. Secondly, to produce livestock with specific genetic and heritable traits. Scientist Ian Wilmut joined Keith Campbell on the project, and their experiment was conducted a staggering 277 times before their work was a success.

A Very Impressive Pair of Glands

The process they used to create Dolly is called somatic cell nuclear transfer. It involved using three mothers: one provided an egg cell, the second the DNA, and the third carried the embryo to term. Using microscopic needles, the nucleus from an egg cell belonging to sheep one was carefully removed, resulting in an empty egg cell ready for implantation. Next, the nucleus from the somatic egg cell belonging to sheep two was extracted and kept (the somatic egg cell contained the DNA required for Dolly).

Finally, the nucleus from the somatic egg cell was fused with the empty egg cell, which was then implanted into the womb of sheep three. 148 days later, Dolly, a Finn Dorset sheep was born, an exact genetic duplicate of the second sheep used for her genesis. She was named after the singer Dolly Parton.

The somatic egg cell was taken from the mammary gland of the second sheep, and when Ian Wilmut was questioned about the origin of her name, he said, 'Dolly is derived from a mammary gland cell and we couldn't think of a more impressive pair of glands than Dolly Parton's.'

The Art of the Put Down

Dolly's life was spent at the Roslin Institute, where she produced six uncloned offspring: Bonnie, Sally and Rosie (twins) and Lucy, Darcy and Cotton (triplets). Sadly, on 14 February 2003, Dolly had to be euthanized. She suffered from crippling arthritis and a lung cancer called *jaagsiekte*, a fairly common disease amongst older sheep, especially those kept inside. Her lung disease was severe and would not clear up, so after a veterinary examination it was recommended she be painlessly 'put down'.

She died at the age of six, which was half her life expectancy. The scientists at the Roslin Institute claimed there was no connection between her premature death and her being a clone, but elsewhere professionals in the field debated this claim. There was speculation that arthritis is usually found in older sheep, and this caused scientists to re-examine how old Dolly really was.

One theory was that as Dolly's genetic mother was six years old when her DNA was taken, this could mean that the 'true age' of clones is their age since birth, plus the age of the genetic donor. Others believed that age is simply determined from the moment of birth. If this logic is followed then Dolly died young, leading to concerns that there may be a risk of premature aging (and diseases associated with aging) in clones. Debates erupted surrounding what could happen if the science developed and a human baby was cloned. In-keeping with the strange life she led, a taxidermied Dolly is on display in the Museum of Scotland in Edinburgh.

Stem Cell Genetics

Whether Dolly the sheep is celebrated or feared, there can be no doubts that her existence was beneficial to the field of biology. The lessons learnt from Dolly can help in the treatment of certain illnesses through the use of stem cells. In conditions associated with damaged cells which cannot repair themselves, such as Parkinson's, diabetes, arthritis, liver damage and macular degeneration, new cells can be introduced into damaged tissue, to treat disease.

The genetic technique involves taking cells from the disease sufferer, and using the genetic material inside the cells to clone an embryo. The stem cells contained within the embryo can progress to become different types of cells, each with a different function. The stem cells are grown *in vitro* and then inserted into the body to replace the lost or damaged cells.

However, there are many ethical implications in this controversial method of treatment, and it is even illegal in some countries due to its use, destruction and creation of human embryos. There are some types of stem cell treatments that do not require the use of an embryo, and a less-controversial method has been pioneered in Japan, using skin fragments to grow stem cells instead.

Genetically modified animals can be used for medical and therapeutic purposes. At the Roslin Institute, cloned sheep modified to produce a certain protein in their milk have been engineered. This milk, which helps the blood to clot, could eventually be used in the treatment of haemophiliacs, people who lack protein and can lose life-endangering quantities of blood if even slightly wounded. The organs of cloned animals can also be engineered for use in human transplants; yet again, this raises ethical questions, and is strongly opposed by animal rights activists.

Reviving the Extinct Species of the World

Scientists argue that cloning may have applications in the future if we wish to protect endangered species from extinction. By preserving frozen tissue, this can be used to ensure the continued survival of certain animals.

Theoretically, it could be conceivable to revive species that are already extinct, if the genetic material was available. In January 2009, a Pyrenean ibex which had been formally declared extinct in 2000, was resurrected by scientists in northern Spain. The newborn ibex did not live for more than a few minutes, but its existence, though short, heralded another huge advance in the field of genetic technology.

SPACE INVADER

Wernher von Braun (1912 – 1977). **German rocket scientist, and space architect, Wernher von Braun, developed the V2 rocket during World War II, a technology used to bomb Britain. Fast-forward 15 years and von Braun was appointed director of NASA, becoming the man responsible for sending American astronauts to the moon.**

Dreaming of the Cosmos

Wernher von Braun was born into an aristocratic family on 23 March 1912 in Wirsitz, Germany. When World War I ended, Wirsitz was transferred to Poland, the von Braun family moved back to Germany. By the time young Wernher was 12 years old he had developed a keen interest in rocket-propelled cars and speed records. One day he attached fireworks to his toy wagon and set it off in a crowded Berlin street. His father collected him from the police station later that day.

Wernher Von Braun grew up reading the science fiction writings of Jules Verne and H. G. Wells, and dreamt of one day going into space. His mother had bought him a telescope, and he spent hours staring at the stars and pondering the possibilities of space flight. In 1930 he enrolled at the Technical University of Berlin, where he worked on military rockets, however, he remained fiercely interested in space flight above all else.

Developing the V2 Ballistic Missile

While at university, von Braun joined the *Verein fur Raumschiffarht (VfR)*, the Spaceflight Society, an amateur German rocket association which was founded in 1927. The group, dubbed the 'Rocket Team', would conduct liquid-fuelled rocket motor tests and design rocket technology.

Von Braun remained fixated with building large rockets, and in 1932 he was employed by the German army to design ballistic missiles. While working on the missiles, he completed a PhD in physics. In 1937, the German army opened Peenemünde, a facility on the north coast of Germany. Von Braun was employed as technical director, and his *VfR* group worked there alongside him, developing rockets and other advanced forms of weaponry.

At the start of the war the Rocket Team developed the design for the A4 (later named V2) ballistic missile, but Adolf Hitler was not impressed. The team kept working on this, however, until Hitler eventually fully supported the construction of this 'wonder weapon'. The V2s were needed quickly and so production was moved to an underground forced labour camp Mittelwerk, populated by 9,000 prisoners from the concentration camp, Mittelbau-Dora. 2,541 workers lost their lives during their time at Mittelwerk, and 5,923 were badly injured.

In March 1944 von Braun was briefly held by the Gestapo on suspicion of treason. They claimed he was more interested in his personal goal of going into space than he was with developing an effective weapon of war. From this moment on and until the end of his life, von Braun used this episode to claim he was a victim of the Nazi regime rather than an accomplice. However, there are testimonies that support the claim that von Braun was involved in activities at Mittelwerk, and that he not only observed the inhumane treatment of the prisoners, but positively encouraged it.

Moon Travel Plan

In September 1945, von Braun surrendered to the Americans, along with information he had gathered over the years. Von Braun and his Rocket Team were brought to the United States under contract with the US Army Ordnance Corps, as part of a top-secret military operation called Project Paperclip. This involved the post-war recruitment of top Nazi scientists by the US government, for their expertise and knowledge. Von Braun spoke out about his surrender to the Americans:

We knew that we had created a new means of warfare, and the question as to what nation we were willing to entrust this brainchild of ours was a moral decision more than anything else. We wanted to see the world spared another conflict such as Germany had just been through, and we felt that only by surrendering such a weapon to people who are guided by the Bible could such an assurance to the world be best secured.

Dr Wernher von Braun at his desk with a moon lander painting in the background and rocket models on his desk.

Von Braun and his Rocket Team were stationed at Fort Bliss, Texas. There they developed ballistic missiles and developed rockets for different purposes; firstly to carry nuclear weapons, then satellites and eventually people. When the Soviet satellite *Sputnik* was launched in 1957, President John F. Kennedy decided the race to the moon was on, finally giving von Braun the opportunity to make his dreams come true.

On 1 July 1960, the German scientists were transferred to the newly-established National Aeronautics and Space Administration (NASA) where von Braun was made director. Von Braun's most famous achievement during his time at NASA was building the giant Saturn V rocket, the super-booster that on 16 July 1969 launched the crew of *Apollo* 11 into space, and allowed them to set foot on the moon. Von Braun's lifelong ambition had been realized.

Secure in the Annals of Space Flight

Wernher von Braun retired from NASA in 1972, and in the following year was diagnosed with cancer. He died a naturalized American citizen on 16 June 1977 in Virginia, United States. Apollo Space program director Sam Phillips was quoted as saying that he did not think that America would have reached the moon as quickly as it did without von Braun's help. Later, after discussing it with colleagues, he amended this to say that he did not believe America would have reached the moon at all.

It is a strange legacy von Braun has left behind. To some he is the ultimate architect of space exploration, and the reason America got to the moon. But his haunting work within Nazi Germany remain forever. In the end, he may have been just a visionary aerospace engineer, caught up in the most repulsive of times doing what he was told.

However he is judged, Wernher von Braun's place in the annals of space flight is secure – the von Braun crater on the Moon is named after him.

Saturn rocket launch, Florida, 29 January 1964. Developed at MSFC under the direction of Dr Wernher von Braun, the SA-5 incorporated a Saturn I, Block II engine and was the first two stage rocket with orbital capability.

A squirrel monkey named Baker peers out from a NASA biocapsule as she's readied for her first space mission. Baker and a rhesus monkey named Able were fired 300 miles into space in the nose-cone of a Jupiter missile AM-18 from Cape Canaveral in Florida on 28 May 1959. The flight, which reached speeds of up to 10,000 mph (16,090 kmh) lasted 15 minutes and the monkeys were recovered 1,500 miles (2,413 km) away in the South Atlantic near Puerto Rico. The pair became the first living creatures to survive a space flight. Miss Baker, as she came to be known, spent the latter part of her life at the US Space and Rocket Center in Huntsville, Alabama. She died of kidney failure in 1984 at the ripe old age of 27.

Launch of a Jupiter missile from Cape Canaveral, Florida, USA. Jupiter missiles were used in a series of suborbital biological test flights. On 13 December 1958, Jupiter AM-13 was launched with a Navy-trained South American squirrel monkey named Gordo onboard. Unfortunately, the nose cone recovery parachute failed to operate and Gordo did not survive the flight. The nose cone sank in the Atlantic 1,302 nautical miles south of Cape Canaveral and was not recovered.

PAGAN ROCKET MAN

Jack Parsons (1914 – 1952). There were two very different sides to Jack Parsons. He was a new-age rocket scientist, seeking to start humanity's space age and an old-style occultist chanting to the Greek god Pan before rocket launches. Parsons saw no contradiction between the ritual magic and rocketry that ruled his world.

Suicide Squad

Marvel Whiteside Parsons was born on 2 October 1914, and is better known as Jack Parsons. He had an unhappy childhood, in a rich but broken home. The Parsons family lived on the exclusive 'Millionaire Row' in an affluent neighbourhood of Pasadena, California. His parents' marriage was an unhappy one, and when they divorced, the young Jack was affected deeply. He had a hatred of authority, and would lash out if anyone interfered with his personal life.

At age 13 he began to take an interest in science fiction and the occult, and once claimed to have summoned Satan. He had no formal education post-college and after being employed briefly at an explosive powder company, he started working at Caltech, the California Institute of Technology. He worked there as a rocket propulsion researcher and Jack and his friends would terrorize the campus with their rocket experiments.

The gang were named locally as the Suicide Squad, and Caltech despaired of their reckless hobby. However, when World War II broke out, the authorities turned their attention to the young misfits and their unique knowledge of explosives. The US military needed the Caltech scientists to design a method of propelling planes into the air in areas which did not have long enough runways.

Through this work, Parsons and his fellow Suicide Squad members formed the Jet Propulsion Laboratory (JPL), which went on to work alongside NASA, constructing and operating robotic planetary spacecraft, as well as conducting Earth-orbit missions. Despite Parsons' lack of education, his aptitude for science and chemistry was clear.

He designed the chemical composition of liquid rocket fuel, and invented JATO (jet-assisted take-off) which would make sending rockets into space possible. Later, Parsons and his colleagues started a company called Aerojet Engineering, which manufactured and sold rockets. However, it was Parsons' dual interest in both inner and outer space which lead him to abandon the venture.

Ordo Templi Orientis

In 1941 Parsons became enraptured with the writings of British ceremonial magician and occultist Aleister Crowley. In 1904, Crowley was in Egypt when he had a religious experience. It dawned on him that he was the prophet of a new age called the Aeon of Horus, and from this realization he developed and established his new religious philosophy, Thelema. Then, a non-corporeal entity approached him. This being, calling itself Aiwass, approached Crowley and dictated a prophetic text called *The Book of the Law*, which included the principles of Thelema.

Jack Parsons came to the attention of Crowley through his reputation as an eccentric rocket scientist, and he was then personally invited by Crowley to join his magical order, Ordo Templi Orientis (O.T.O). This was based at Agape Lodge in Los Angeles, and Parsons was heralded as the successor to the 'Great Beast' himself.

Crowley's colleagues at O.T.O. saw him as a potential saviour for their movement, and he began donating his salary to the brethren. In 1942, Parsons' father died, and bequeathed to him a sprawling mansion in Pasadena. Parsons moved in and it soon became a centre for science fiction and occult discussions, as well as attracting California's bohemian population.

In 1945, L. Ron Hubbard, the controversial figure who went on to found the Church of Scientology, befriended Parsons and

The Great Beast, Aleister Crowley (1875–1947), English author, poet, prophet, occultist magician and leader of Ordo Templi Orientis was one of the most controversial men of his time. Photographed in 1955.

moved into the mansion. The pair became 'magical partners' for some time, much to the disgust of Crowley, who – despite the fact he had never met Hubbard – considered him a charlatan.

Babalon Working

Over several nights in the February and March of 1946, Hubbard and Parsons performed a ritual known as 'Babalon Working'. The ceremony was performed by Parsons, and was designed to invoke the 'Scarlet Woman', also known as Babalon. Hubbard observed the ritual and made notes.

A few days later, Parsons returned home to discover a woman named Marjorie Cameron was waiting for him. He assumed she was the Scarlet Woman he had conjured, and she confirmed that she was. He wrote of his success to Crowley immediately.

The next step was to attempt to conceive a Moonchild, the messiah of Thelema, through the erotic and bizarre ritual of 'sex magic'. The Moonchild was to signal the end of the Christian era, eradicating its outmoded morality and ushering in a new age of free love. Parsons had learnt of this idea from Crowley's 1917 novel *Moonchild*. No child was born, but Parsons still declared the mission a success, as he felt a spiritual entity had been produced.

L Ron Makes a Getaway

After his success summoning the Scarlet Woman, Parsons was faced with betrayal by his friend, L. Ron Hubbard, and mistress, Sarah Northrup. Hubbard proposed a business idea involving buying and selling boats. This venture required almost all of Parsons' savings. Little did he know, Hubbard was planning on fleeing the country on a boat with Sarah. He lost all his savings and the mansion was repossessed. Crowley had predicted this outcome prior to it happening, and spoke out in disgust when his prophecy was fulfilled. Parsons, however, attempted to conjure up a typhoon in revenge.

Dark Side of the Moon

Towards the end of the 1940s, Parsons was caught up in a scandal involving classified documents falling into the wrong hands, and this resulted in the end of his career. He wound up producing minor special effects for Hollywood movies, any dreams of sending man into outer space were gone.

On 17 June 1952 there was an explosion of fulminate of mercury in his backyard laboratory in Pasadena, California. He survived the initial blast, but his injuries proved fatal soon after. When his mother was informed, she killed herself the same day.

Many have speculated that the circumstances surrounding his death were suspicious. Allegedly the FBI were involved, and the 'accident' was actually a sinister plot realized by old enemies. Whatever the truth may be, Parsons died surrounded by magic and chemicals.

Years later, French scientists affectionately named a crater on the dark side of the moon after the legend of aerospace engineering, Jack Parsons.

THE ATOM-SMASHING TIME MACHINE

The Large Hadron Collider (LHC) is the highest-energy particle accelerator in the world. This global experiment of epic importance aims to investigate one of the fundamental forces of nature and also help us to understand how the early universe evolved. In Russia, however, two physicists Irina Aref'eva and Igor Volovich thought the LHC could be used as something else: a time machine.

Colliding with the Hadrons

The LHC was built at the European Organization for Nuclear Research (CERN) in Geneva and is located 100 m under the French-Swiss border. It is worked on by 1000 scientists from 33 countries and is considered to be one of the most important experiments of the 21st century.

The LHC is 27 km long and forms a circular tunnel made primarily of super-conducting magnets. The collider works by accelerating hadrons (either protons or lead ions) to nearly the speed of light and exposing them to temperatures 500,000 times hotter than the core of the sun. The hadrons smash together. When they collide the hadrons create sub-atomic fireballs with incredible temperatures and densities.

The collider generates around 10,000 particle collisions per second. In this environment, the LHC simulates the same cosmic conditions of a millionth of a second after the Big Bang.

Physicists believe that everything began approximately 13.7 billion years ago when the Big Bang gave birth to the universe. At this point in time the universe was hot and dense, and as it cooled 'everything' started to develop. With the help of the LHC, the mysterious beginnings of our universe can potentially be unravelled.

The Mysteries of the Universe

The universe is unfathomably big, and what can be observed by conventional methods is estimated to only be 4% of what's out there. As the remaining 96% does not emit any light, we can't locate it; but we know it is there because of its effects.

The celestial bodies we can see respond to gravitational forces which are stronger than the ones we know about, and in addition to this, the universe's expansion is accelerated by an unknown force. These undiscovered forces are labelled dark matter and dark energy.

Dark matter is believed to make up 26 per cent of the universe, and dark energy accounts for the remaining 70 per cent. These two 'substances' do not consist of particles of the standard model, and so physicists hope that the LHC collisions will produce the unknown particles dark matter is made of, and make them detectable.

While the LHC makes huge and important leaps in the field of physics, Professor Irina Aref'eva and Dr Igor Volovich, mathematical physicists at the Steklov Mathematical Institute in Moscow, believe it could potentially become the first ever time machine.

Wormholes in time and space

When the LHC was first switched on at CERN in 2008, Aref'eva and Volovich claimed we could expect visitors from the future to arrive at the same time. The physicists argued that the hadrons have the same kinetic energy as a flying mosquito, but when this energy is concentrated into a sub-atomic particle, one trillionth of a mosquito's size, the immense impact of this collision can affect 'space-time', a term used to describe the fabric of the universe.

Space-time can distort into loops known to science fiction fans as 'wormholes'. These wormholes are like tunnels linking different parts of space and time together. Theoretically they open passages (or portals) to receive visitors from another time or planet.

The physicists claim the LHC is capable of creating such wormholes, something which would prove their existence.

However, scientists at CERN argue that even if wormholes were created, their dimensions would be tiny, only allowing sub-atomic particles to pass through. Neither is it known how a wormhole would stay open – closure could prove fatal to time travellers.

Aref'eva and Volovich have speculated that dark energy would be powerful enough to keep a wormhole open. In theory, wormholes could allow us to receive people, objects or information from other times and galaxies. Some are amazed by this intriguing possibility while others argue that this would make us vulnerable, with no control over what could be transported.

There is no evidence to support the idea that wormholes exist, but Aref'eva and Volovich have worked it out. If there is less energy in the LHC after a collision, then it can be assumed some particles have escaped through a wormhole.

Aref'eva and Volovich also claim that the year the machine was first switched on, 2008, would become 'year zero' in time travel, as it would only be possible to travel back as far as the creation of the first wormhole.

Particle collisions in the Large Hadron Collider. Neutrinos are so elusive they have been dubbed 'ghost particles'.

What if particles really can exceed the speed of light?

The Large Hadron Collider at CERN has enormous underground detectors to catch sub-atomic particles called neutrinos that exceed the universe's speed limit. Perhaps the most exciting thing is that time travel could be feasible via 'wormholes' as popularized in science fiction, connecting one place in space to another vastly distant one.

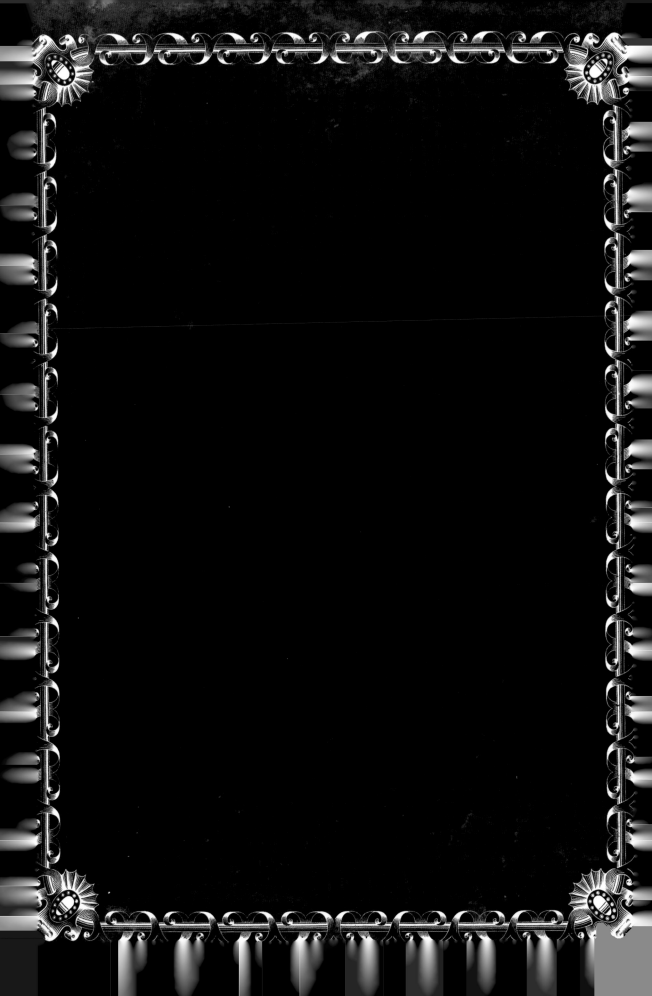

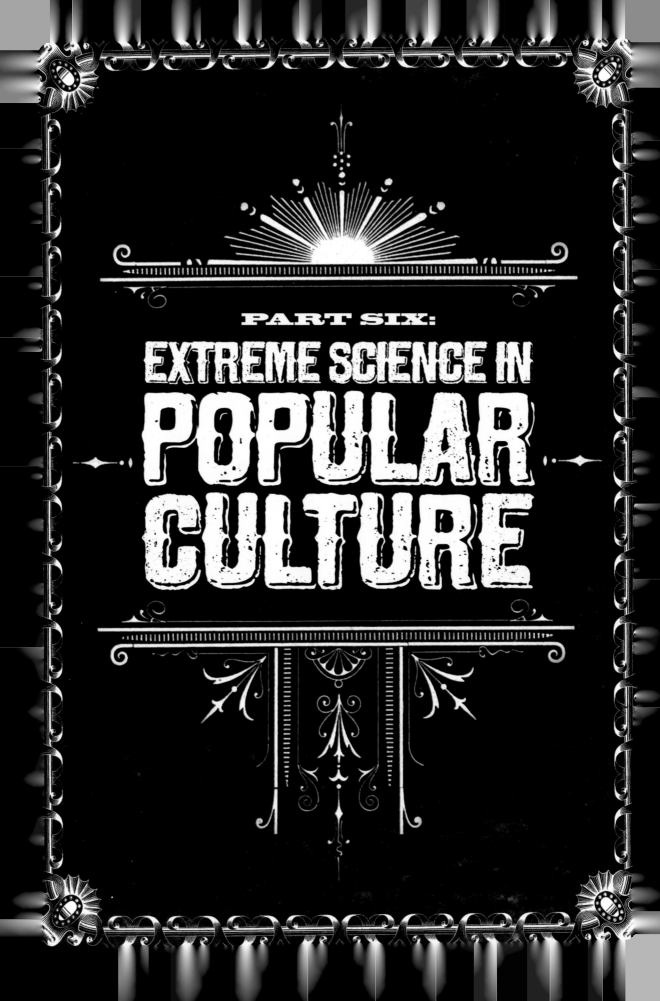

PART SIX:
EXTREME SCIENCE IN POPULAR CULTURE

RE-ANIMATION

FRANKENSTEIN.

"By the glimmer of the half-extinguished
light, I saw the dull, yellow eye of the
creature open; it breathed hard, and a
convulsive motion agitated its limbs.
*** I rushed out of the room."

Page 43.

ondon, Published by H. Colburn and R. Bentley, 1831.

Frankenstein; or, The Modern Prometheus

BY MARY SHELLEY (1797 – 1851)

Frankenstein is a novel written in 1818 by Mary Shelley, wife of the poet Percy Bysshe Shelley. With fused elements of Gothic and Romance, the novel is also considered by critics to be one of the most influential science fiction novels of all time. A major inspiration for subsequent literature and popular culture, it has spawned a complete genre of horror and science fiction stories and movies.

Mary Shelley started writing the story when she was 18, and the novel was published when she was 21. The first edition was published anonymously in London in 1818. Mary Shelley's name appears on the second edition, published in France in 1823.

The Shelleys had travelled widely in France accompanied by Lord Byron and writer John Polidori. Galvanism was the controversial topic of the day, and sparks began to fly whenever the subject was raised in conversation among the group of travelling friends. Byron challenged Mary to write a ghost story and the result was her Frankenstein novel. Allegedly coming to her in a dream, the plot is inspired by Giovanni Aldini's electrical galvanism experiments on dead animals (see p. 24).

Frontispiece illustration from Mary Shelley's novel, *Frankenstein; or, The Modern Prometheus*, first published 1818.

Obsessed with Science

In the novel, Doctor Victor Frankenstein, a man obsessed with science, galvanism and chemistry, decides to create a living person artificially. It is not clear from what material he makes this person, but the reader gets a sense that he uses body parts of humans and animals. The techniques used to bring his 'creature' to life are ambiguous too, though with Frankenstein's interest in ceremonial magic and galvanism it is assumed he uses these secret methods to animate the body.

When the process is finished, Dr Frankenstein is horrified at the monster he has created, and hoping to simply forget about it, he flees and attempts to lead a normal life elsewhere. The monster, however, is confused and angry at being abandoned, and wreaks havoc on Dr Frankenstein's life in revenge for his rejection.

Shelley's work expresses her fear of man's evil capabilities given the right technology. When the book was written, the process of galvanism had just been discovered. In galvanism, electrical current is used to re-animate matter such as the lifeless corpses, with the intent of bringing the dead back to life.

The 1931 Universal Pictures adaptation of the Frankenstein story is probably the most famous and celebrated of all the versions. Starring Colin Clive as Dr Frankenstein and Boris Karloff as the monster it features many differences from Shelley's original text. The monster is portrayed very differently in the film. He is quieter, expressing himself in mostly grunts, compared to the novel where he can speak articulately.

The Monster is Resurrected

The story's theme is re-animation and the monster came back to life with a vengeance in the movie. The Universal make-up department went to town and created the unforgettable bolt in the neck look. In the film, the monster is the recipient of a criminal brain, a mind already contaminated with dark thoughts and violent tendencies. Whereas, in the novel, the monster's behaviour is due to the neglect and rejection of his creator, Dr Frankenstein.

The movie also added a cast member, the Doctor's assistant, Fritz (Dwight Frye). This extra personality contributed a new dimension to Dr Frankenstein's character which is not in-keeping with the original novel. In Shelley's book, the Doctor was a solitary man of science, and always worked alone. In the movie his thoughts and re-animation deeds are shared with Fritz. (The assistant's name changed to Igor in the 1939 movie *Son of Frankenstein*, a name deemed more suitable for the evil hunchback character played by Bela Lugosi.

The 1931 film spectacularly shows the method of re-animating the body with electricity, something which the novel skirts around. In the movie, Doctor Frankenstein is shown searching for recently dug graves and exhuming corpses. It is clear his human project will be constructed from organs and deceased body parts.

Shocking Tour de Force

The dazzling dramatic creation scene is a remarkable *tour de force* even by today's standards. The laboratory is filled with science apparatus, flashing light bulbs and shining instruments. The patchwork body lies on a table and as lightning can be heard, Doctor Frankenstein and Fritz begin excitedly moving around the laboratory, turning wheels and flicking switches. They peel back the covers from the body and elevate the table it lies on, moving it closer to the ceiling and through the roof, to allow it to receive the charge of the lightning bolt.

As lightning strikes above them, Doctor Frankenstein and Fritz stare up in anticipation. The body is lowered back down to the laboratory, and they watch in amazement as its fingers begin to move, and its arm raises upwards. 'It's alive!'.

The name Frankenstein is often incorrectly used to refer to the monster itself. Part of Dr Frankenstein's rejection of his creation is the fact that he does not give it a name, which gives it a lack of identity. In the novel, the monster is identified via words such as creature, monster, fiend and wretch. Speaking to Dr Frankenstein, the monster sorrowfully refers to himself as 'someone who would have been your Adam, but is instead your fallen angel'.

SCIENCE'S MONSTER TERROR !

BORIS KARLOFF

X CERT

FRANKENSTEIN

Poster for the 1931 movie *Frankenstein*.
Directed by James Whale and starring Boris Karloff.

FRANKENSTEIN INSPIRED MOVIES

Frankenstein

DIRECTED BY JAMES WHALE (1931)
STARRING COLIN CLIVE, BORIS KARLOFF,
MAE CLARKE

A horror classic in which obsessed scientist, Henry Frankenstein (named Victor in the novel), assembles a living being from parts of exhumed corpses.

To the seen-it-all eyes of the 21st century viewer, the 1931 movie *Frankenstein* doesn't seem so scary. Overexposure has made the present-day audience immune to horror and violence. But to the innocent movie-goers of the early 1930s it was the most frightening film they had ever seen.

Maybe the film doesn't give us a cold shiver anymore, but it is still a great movie. The scene in which Colin Clive as Dr Frankenstein shouts hysterically 'It's alive! It's alive!' is recognized as one of the greatest ever moments of horror film history.

After bringing the monster to life, Dr Frankenstein utters the equally infamous line, 'Now I know what it's like to BE God!' The movie was originally released with this line of dialogue, but when it was re-released in the late 1930s, censors demanded it be removed on the grounds that it was blasphemous. A loud clap of thunder was substituted on the soundtrack.

Boris Karloff's portrayal of Frankenstein's creature may not be too scary to the modern watcher, but his performance is truly phenomenal and his iconic monster make-up has made him immortal. *Frankenstein* is the ultimate monster classic of Universal studios

and Boris Karloff's role as the creature of Frankenstein is the most recognized and beloved monster of all time.

The Bride of Frankenstein

DIRECTED BY JAMES WHALE (1935)
STARRING BORIS KARLOFF, ELSA LANCHESTER
AND COLIN CLIVE

Director James Whale originally did not want to do a sequel to *Frankenstein* and Universal Studios considered producing it without him. However, after 4 years of nagging by studio bosses, Whale relented.

Dr Frankenstein and his monster are both still alive, not killed as previously believed. Frankenstein wants to quit the evil experiment business, but when mad scientist, Dr Pretorius, kidnaps his wife, he is blackmailed into creating a new female companion for the creature: The Bride of Frankenstein.

Few sequels are deemed superior to their predecessors, however, critics say that *The Bride of Frankenstein* not only equals its prototype *Frankenstein* (1931), but surpasses it. Despite the first film's reputation, the sequel's combination of horror and wicked humour make it a more enjoyable watch. Needless to say, both films are justly hailed as classics.

Elsa Lanchester plays both the parts of writer Mary Shelley and the eponymous heroine, the monster's bride. She was only 5ft 4in tall but wore stilts that made her 7 feet tall for the part of The Bride. One of the most obscure of the classic movie monsters The Bride is only on screen for five minutes and never kills anyone. The movie's line *'We belong dead'* was voted as number 63 of The 100 Greatest Movie Lines by *Premiere* magazine.

CREATURE WITH THE ATOM BRAIN

SHOCK-FULL OF THRILLS!

RICHARD

with **ANGELA S**

Story and Screen Play by **CURT SIODMAK**

A CLOVER PRODUCTION ·

Creature with the Atom Brain was a 1955 B-movie zombie film, directed by Edward L. Cahn from a screenplay by Curt Siodnak and distributed by Columbia Pictures as the bottom half of a double bill with *It Came from Beneath the Sea*. The cast included Michael Granger, Gregory Gaye as well as Richard Denning, who starred in a number of similar 1950s B-movies. The plot was an extreme science classic: a deported American gangster Frank Buchanan (Michael Granger) forces ex-Nazi scientist Wilhelm Steigg (Gregory Gaye) to create killer zombies by resurrecting corpses with nuclear radiation to exact revenge on his enemies.

Son of Frankenstein

DIRECTOR ROWLAND V LEE (1939)
STARRING BORIS KARLOFF, BASIL RATHBONE,
BELA LUGOSI

In the late 1930s, Universal Studios decided it was time to resurrect the Frankenstein Monster one more time. *Son of Frankenstein* is one of the most visually impressive of all of Universal's horror films. Transylvania is a world of perpetual fog, rain-swept castles and blasted heaths with terrifying lightning storms. Shadowy corridors lead to the Frankenstein crypt, in which both grandfather and father are dead, but the step-brother, the monster and family black sheep is very much alive – *We're all dead here!*

Wolf Frankenstein, son of Henry Frankenstein, claims his inheritance. While exploring his father's laboratory he encounters the hunchback Igor, who begs him to revive his father's comatose monster. Wolf fails to revive the monster. But following a spate of local murders, the villagers immediately connect the killings to Frankenstein. The investigating police inspector discovers that the monster is alive after all and being manipulated as an instrument of death by the evil Igor, who Wolf ends up shooting. The enraged monster goes berserk after losing his only friend. In the finale, Wolf tracks the monster back to the lab and throws him into a sulphur pit to his apparent demise.

The film may have the greatest horror film cast ever. Boris Karloff dominates as the Monster, but Bela Lugosi's performance as crooked murderous hunchback Igor is considered by many critics to be his greatest.

Abbott and Costello Meet Frankenstein

DIRECTOR CHARLES BARTON (1948)
STARRING BUD ABBOTT, LOU COSTELLO,
LON CHANEY JR, BELA LUGOSI

A favourite Abbott & Costello comedy, this movie is both a spoof and an homage to the legendary movie monsters of Universal. The dark, Gothic sets are full of flapping bats, cobwebs and evil laboratories. Lon Chaney Jr. is the Wolf Man Larry Talbot, Bela Lugosi is Dracula and Glenn Strange is the Frankenstein Monster as Abbott and Costello's slapstick erupts around them.

A bad moon rising in fog-bound London town. Count Dracula and the Frankenstein Monster have been shipped to a wax museum. When the sun sets, Dracula and the Monster will rise again. Larry Talbot, the Wolf Man, knows he must warn someone but it's Lou Costello who answers the call. The fun begins. Dracula needs a new brain for his monster, and Lou Costello is the perfect subject. Bud Abbott and The Wolf Man have to stop Dracula before Lou loses his head!

Young Frankenstein

DIRECTOR MEL BROOKS (1974)
STARRING GENE WILDER, MARTY FELDMAN,
PETER BOYLE, MADELINE KAHN

The screenplay is written by Mel Brooks and Gene Wilder who also stars as Dr Frederick Frankenstein, the grandson of Victor von Frankenstein an obscure relative of the original Dr Frankenstein. The comedy is an affectionate parody of classic horror movies of the past, and replicates the style of the original 1931 *Frankenstein* movie by filming entirely in black and white. The hunch-backed, bulging-eyed assistant, Igor, is played memorably by Marty Feldman with the monster played by Peter Boyle. Most of the lab equipment props were created by Kenneth Strickfaden for the original Universal classic.

Frederick Frankenstein is a respected lecturer at an American medical school but has always disowned his notorious grand-father. Until, that is, he inherits the family castle and travels to Transylvania. Though the Frankenstein legend has brought nothing but shame and ridicule, Frederick becomes increasingly intrigued by his grandfather's corpse re-animation work. When he discovers his grandfather's lab and reads his scientific notes, Frederick becomes a man possessed and resumes his grandfather's experiments in re-animating the dead. Robbing a corpse from a recent grave, Frederick and Igor get to work. Soon, Young Frankenstein is ready to re-animate his very own creature...

OTHER WORKS INSPIRED BY FRANKENSTEIN

Weird Science

DIRECTED BY JOHN HUGHES (1985)
STARRING ANTHONY MICHAEL HALL,
ILAN MITCHELL-SMITH, KELLY LeBROCK.
In this cult teen comedy, inspired by the Frankenstein story, two teenage nerds, Gary and Wyatt, are beating themselves up over being useless with girls. They decide to create a computerized 'perfect virtual woman'. She must be gorgeous, to impress the school bullies, but she must also be intelligent and

understanding, so they can converse with her and she can teach them how to be suc-cessful with girls. She must also, of course, be devoted to their every need.

Their computer looks like a cluttered experimental science lab, but they set to work programming their perfect woman. Taking photos and vital statistics from glamour magazines they specify the exact dimensions of her body and every little de-tail of her personality including making her fingers play piano. When their computer reaches maximum capacity, they realize they need more power and so hack into the government mainframe, sending the neighbourhood's power supply into chaos. In the experiment's climax, they connect the computer to a Barbie doll (who happens to be positioned on top of the board game 'Life').

When they hit *ENTER* to execute their program, a lightning bolt strikes the house leading to a violent explosion. When the smoke clears a beautiful woman, Kelly Le Brock, emerges. Interrupted by a snippet from the 1935 movie *Bride of Frankenstein* with Victor Frankenstein shrieking, *'She's Alive, Alive!'*, in stark contrast their sexy creature's first words to the boys are 'So what would you little maniacs like to do first?'

Re-Animator

DIRECTOR STUART GORDON (1985)
FROM A STORY BY H.P. LOVECRAFT
STARRING JEFFREY COMBS, BRUCE ABBOT,
BARBARA CRAMPTON
Professor Herbert West is a scientist who has discovered a formula which brings the dead back to life by re-animating their tis-

sue. The film is a horror comedy and, once re-animated, the walking dead create plenty of havoc, blood and splattered brains.

Re-animator was made on a budget of less than a million dollars, which is truly remarkable for a movie containing so many special effects. It seems that necessity may have been the mother of invention with practical low-cost solutions replacing expensive effects. The 'brains' in the smashing severed heads scenes were made up of meat by-products and fake blood. The crew wore garbage bags over their clothes. No one knew just how far brains splatter!

RE-ANIMATOR SEQUELS
II. Bride of Re-Animator (1990)
III. Beyond Re-Animator (2003)

Many sequels, particularly horror sequels just recap the original with more gore, but the *Re-Animator* sequels haven't. In *Bride of Re-Animator*, a new idea about animating different parts of a dead corpse emerged, and attempts were made to add the 'soul' to the re-animated body. *Beyond Re-Animator* features an exploding chest and severed torso that walks on its hands. They are both underrated, if incredibly gory, sequels.

Hammer Film Productions

A film production company based in London, Hammer was founded in 1934, and is renowned for a series of Hammer Horror movies which were at their zenith in the 1950s and 1960s.

Directed by Terence Fisher *The Curse of Frankenstein* was Hammer's first smash hit in 1957, not only in Britain but also in the USA. Bringing Gothic horror alive with Hammer's new colour technology, Fisher incorporated unprecedented amounts of gore and explicit sexual overtones. The huge box office success of *The Curse of Frankenstein* led to the inevitable sequel and *The Revenge of Frankenstein* was released in 1958. Hammer consolidated their ascendancy by producing another five Frankenstein horror movies over the next 15 years. Perpetuating the Monster's myth for another generation of movie-goers, they made leading stars of British actors Christopher Lee and Peter Cushing.

THE HAMMER FRANKENSTEIN MOVIES

The Curse of Frankenstein (1957), *The Revenge of Frankenstein* (1958), *The Evil of Frankenstein* (1964), *Frankenstein Created Woman* (1967), *Frankenstein Must Be Destroyed* (1969), *The Horror of Frankenstein* (1970), *Frankenstein and the Monster from Hell* (1974).

Re-animation scene from *Beyond Re-Animator* (2003). Director Brian Yuzna.

TIME TRAVEL

The Time Machine

Novel by H.G. Wells (1866 – 1946)

A pivotal science fiction story by H.G. Wells was published in 1895. *The Time Machine*, a hugely influential short book (it is only 32,000 words), has been adapted into at least two feature films and is generally credited with the popularization of the concept of time travel. A 'time machine' is a vehicle that allows the 'time traveller' to journey backwards and forwards through time. It is a common device these days, but in 1895, time machines were unheard of.

The book's hero is an English scientist and inventor who is identified simply as the Time Traveller. He has built a machine capable of carrying a person through the fourth dimension of time. The Time Traveller tests his invention with a journey that takes him to the year AD 802,701, where he meets a society of small, elegant, childlike adults – the Eloi.

They live in small peaceful communities doing no work. However, when darkness falls, he meets the evil ape-like troglodytes, the Morlocks, who live underground and rule the night.

In the Morlock's subterranean caves, he discovers the machinery and industry that makes the peaceful community above ground possible. He saves an Eloi female named Weena from drowning and they develop an affectionate relationship. They are then attacked by Morlocks in the night, and

Weena faints. The Traveller escapes in the time machine but Weena gets left behind.

He travels 30 million years further ahead from his own time where he watches the sun grow dimmer, and the world freezing over

Cover of the first edition of *The Time Machine*, novel by H.G. Wells (1895).

as the last living things on Earth die out. Staggered by this vision of the future, he returns to his laboratory, arriving just three hours after he originally left.

The next day, the Time Traveller sets off on another mission presumably to try to rescue Weena. He promises to return in half an hour, but never does.

H.G. Wells' The Time Machine

DIRECTED BY GEORGE PAL (1960)
STARRING ROD TAYLOR, ALAN YOUNG AND YVETTE MIMIEUX

In 1960, an American science fiction film based on the novel was released, featuring the character H. George Wells as the Time

Traveller. In the movie, George knows that his machine travels through time without moving anywhere. It is geographically stationary. The patch of ground occupied by his laboratory would eventually be the place inhabited by the Eloi and Morlocks 800,000 years in the future. Regarded as perennial classic, the special effects were way ahead of their time. The film received an Oscar for the time-lapse photographic effects showing the world changing rapidly.

George Pal also filmed H.G. Wells' *The War of the Worlds* (1953) and had always intended to make a sequel to his 1960 film, but a second movie was not produced until 2002 when Simon Wells, great-grandson of H.G. Wells, directed an updated revised version of the original story.

Rod Taylor as George Wells in *The Time Machine* (1960). When George arrives in the year 802701 the date on his time machine reads October 12, the same date Columbus first reached America.

OTHER WORKS INSPIRED BY TIME TRAVEL

Doctor Who

BBC TV (1963 – PRESENT)

Inspired by Wells' *Time Machine*, *Doctor Who* is a British science fiction television programme produced by the BBC, depicting the adventures of a mysterious time traveller known only as the Doctor. With great intelligence he battles injustice while exploring time and space in a time machine, the TARDIS (an acronym for Time And Relative Dimension(s) In Space). His time machine appears much larger on the inside than on the outside. It was eventually revealed that the Doctor was a Time Lord and a fugitive from his own people of the planet Gallifrey. With a succession of companions, the Doctor faces a wide variety of alien foes such as the notorious Daleks, while working to save civilizations and right wrongs.

Doctor Who has been critically acclaimed for its imaginative story lines and creative low-budget special effects. The show is a significant part of British popular culture and it has become something of a cult classic in the United Kingdom. *Doctor Who* first appeared on BBC television on 23 November 1963. The first episode was overshadowed by the assassination of US President John F. Kennedy the previous day. So the BBC re-broadcast it the following Saturday. The series' famous theme tune was composed by Ron Grainer of the BBC Radiophonic Workshop.

The Doctor has changed appearance 10 distinct times since the first episode in 1963. The plot line allows for an identity-altering process known as regeneration when the actors playing the Doctor change over. The old Doctor is replaced by a brand new one. Viewers have got used to the change overs and they are as eagerly anticipated as the change over in the James Bond actors. The first Doctor was played by William Hartnell in 1963 – 66. The most recent is Matt Smith who became the Doctor in 2010.

The programme is listed in the Guinness World Records as the longest-running science fiction television show in the world, and is the most successful science fiction series of all time, in terms of its overall broadcast ratings.

The TARDIS. The British Police Box used in the *Doctor Who* TV series.

Police boxes were introduced in Britain in the 1920s. They were wooden, 4 feet square, and contained a small desk, a stool, electric lighting and a heater. A small compartment containing a telephone and first aid kit was accessible from the outside.

The Planet of the Apes

DIRECTOR FRANKLIN J. SCHAFFNER (1968)
STARRING CHARLTON HESTON, RODDY
McDOWALL, KIM HUNTER

Based on the 1963 novel *La Planète des Singes* by Pierre Boulle, the story was translated into English by British writer and translator Xan Fielding and published as *Monkey Planet* (1963).

After their spaceship crashes on a mysterious planet, two US astronauts wander across a desert plain thinking they are alone. However they find that the planet is ruled by intelligent apes. Walking, talking apes, who see themselves as the original beings on the planet and do anything to keep humans oppressed and enslaved. It is the year 3978 AD. The spacecraft had been travelling through space at the speed of light for 2006 years, and due to the anti-aging effects of time dilation, the astronauts are only 18 months older. They are captured and taken to the city of the apes. One of the astronauts undergoes a lobotomy, transforming him into a state of living death. The other finds himself hunted by the apes. The film highlights issues such as animal rights, evolution, class structure, and nuclear war.

Planet of the Apes is the first film in a series of five. The other films are *Beneath the Planet of the Apes* (1970), *Escape from the Planet of the Apes* (1971), *Conquest of the Planet of the Apes* (1972), *Battle for the Planet of the Apes* (1973).

Sleeper

DIRECTOR WOODY ALLEN (1973)
STARRING WOODY ALLEN, DIANE KEATON,
JOHN BECK

Cryogenically frozen after surgery, Miles Monroe wakes up in 2173. The world is ruled by an oppressive tyrant and Miles has been re-animated by a group of rebels fighting to overthrow the government. For reasons that are far too complex to explain, Miles is forced to go on the run disguised as a robot and finds himself falling in love with his new owner. This is the story of their fight against the regime.

The Philadelphia Experiment

DIRECTOR STEWART RAFFILL (1984)
STARRING MICHAEL PARÉ, NANCY ALLEN,
ERIC CHRISTMAS

Reportedly based on an event that took place in 1943, during World War II. While developing a cloaking device to render their naval vessels undetectable to radar, scientists make one of the ships in Philadelphia Harbor disappear. Two of the ship's sailors find themselves thrown forward in time by 40 years to 1984.

Sequel: *The Philadelphia Experiment II* (1993) An updated version of the original involving time portals, black holes and stealth bombers switching between 1993 and 1943.

Back to the Future

DIRECTOR ROBERT ZEMECKIS (1985)
STARRING MICHAEL J. FOX,
CHRISTOPHER LLOYD

Perhaps the most popular of all time travel movies, *Back to the Future* features the exploits of teenager Marty McFly and his mad scientist pal, Doc Brown, whose greatest invention is the DeLorean time machine. The accelerator of the automobile-shaped

'He's the only kid ever to get into trouble before he was born.' Michael J. Fox and Christopher Lloyd in a scene from *Back to the Future* (1985). Director Robert Zemeckis.

time machine is floored and the car zooms off. When the optimum speed of 88 mph is reached, the DeLorean activates the key component to time travel: the flux capacitor. Seconds later the car vanishes, leaving behind a set of blazing smoking tracks and whisking Marty and Doc off to all sorts of time travelling adventures.

Sequels were also released: *Back to the Future II* (1989) and *Back to the Future III* (1990)

Flight of the Navigator

DIRECTOR RANDAL KLEISER (1986)
STARRING JOEY KRAMER, PAUL REUBENS, VERONICA CARTWRIGHT

A 12-year-old boy, is knocked unconscious. Waking up, he thinks only a short time has passed, but discovers eight years have gone by. Strangely he has not aged a day. At the time of the accident, an extra-terrestrial spacecraft crashes into power lines nearby and is investigated by NASA. It is soon revealed that the boy and the spacecraft are linked telepathically. Hearing a voice calling to him telepathically, the boy enters the spacecraft and assumes the role of the navigator.

Quantum Leap

TV SERIES (1989 – 1993)
CREATOR DONALD P. BELLISARIO (1984)
STARRING SCOTT BAKULA, DEAN STOCKWELL

Dr Sam Beckett leads an elite group of scientists into the desert to develop a top secret time travel project known as Quantum Leap. Pressured to prove his theories or lose funding, Beckett experiments on himself ... and vanishes. He wakes up in the past, suffering from partial amnesia and with a telepathic hologram as his only means of communicating with his own time.

Twelve Monkeys

DIRECTOR TERRY GILLIAM (1995)
STARRING BRUCE WILLIS, MADELEINE STOWE, BRAD PITT

In a future world devastated by disease, a convict is sent back in time to gather information about the man-made virus that wiped out the planet.

Primer

DIRECTOR SHANE CARRUTH (2004)
STARRING SHANE CARRUTH, DAVID SULLIVAN, CASEY GOODEN

Four men in a garage have built a small business of error-checking devices. But, they know that there is something more to life than this. Through trial and error they end up building a time machine that puts anything placed inside into its own time loop. They decide to make the device person sized so they can travel back in time.

Déjà Vu

DIRECTOR TONY SCOTT (2006)
STARRING DENZEL WASHINGTON, PAULA PATTON, JIM CAVIEZEL

An ATF agent investigates a terrorist bombing of a ferry in New Orelans and gets attached to a secret FBI unit. Using experimental time warp technology he travels back in time to change destiny by preventing the bombing and the murder of a mysterious dead woman.

Source Code

DIRECTOR DUNCAN JONES (2011)
STARRING JAKE GYLLENHAAL, MICHELLE MONAGHAN, VERA FARMIGA

Colter Stevens' last memory is flying a US Army helicopter in Afghanistan. He wakes up on a train having assumed the identity of another man. 8 minutes later, the train explodes and Stevens finds himself in an experimental time pod. Controlled by scientists, Stevens is sent back in time to find out who bombed the train and avert the disaster.

'What happens if it actually works?'
Shane Carruth and David Sullivan try out their time machine in *Primer* (2004).

GENETICS AND CLONING

Jurassic Park

NOVEL BY MICHAEL CRICHTON (1942 – 2008)
Following in the lurching footsteps of Frankenstein's monster came Michael Crichton's 1990 science fiction novel warning about the dangers of re-animation and manipulation of genetic clones. *Jurassic Park* tells the story of the collapse into violent chaos of an amusement park of genetically-recreated prehistoric dinosaurs who go hunting humans when the park's electric security fences fail.

The novel became a bestseller and Michael Crichton's most famous work. The book became even more famous when, in 1993, Steven Spielberg released a blockbuster movie winning three Oscars and grossing more than US$900 million dollars.

Even though the cloning of the dinosaurs in the story ended in total disaster, their re-animation for the movie was a massive success, and critics regard *Jurassic Park* as a milestone in computer-generated imagery.

Palaeontologist Jack Horner supervised the dinosaur designs to help Spielberg portray them as normal animals of the time rather than monsters. In early designs Horner dismissed studio ideas such as flicking the tongues of the human-hunting Velociraptors, complaining, 'dinosaurs have no way of doing that!' Taking the expert seriously, Spielberg insisted on no tongues.

Comprehensively detailed models of the dinosaurs were made with moulded latex skins fitted over complex robotics. But Spielberg still found the end results clunky and unsatisfactory. So whizz kid animators were called in to create computer-generated dinosaurs.

An initial *Tyrannosaurus rex* sample was approved and more brilliant recreations of prehistoric extinct stars such as Brachiosaurus and Triceratops were to follow. A stampeding herd of Gallimimus was featured in one memorable scene. Legend has it that when producer George Lucas watched the first computer-generated dinosaur demo, his seasoned eyes began to fill up with tears and he reportedly said, 'It was an unforgettable moment in history, like the invention of the light bulb or the first telephone call. Things were never going to be the same.'

The reality is that cloning a complete DNA sequence to create a genetic replica requires a live female egg cell from the same organism. Since no dinosaurs are alive today, cloning them is impossible. But new advances are being made daily in genetic research, so artificially replicating a female egg from recovered DNA may not be too far beyond the bounds of future possibilities one day.

Jurassic Park

DIRECTOR STEVEN SPIELBERG (1993)
STARRING SAM NEILL, LAURA DERN, JEFF
GOLDBLUM, RICHARD ATTENBOROURGH

On the island of Isla Nublar, near Costa Rica, John Hammond (Richard Attenborough) has created an amusement park of cloned dinosaurs called Jurassic Park. About to be opened to tourists from all around the world, palaeontologist Alan Grant (Sam Neill) and his partner, Ellie Sattler (Laura Dern), are invited to the island. On arrival they immediately see fully-grown dinosaurs roaming freely, and are astonished at Hammond's achievements. However, after a tour of the laboratory where the dinosaurs are created, the scientists start to ponder the ethical and moral issues connected to resurrecting an extinct species. On a tour of the park a security breach causes all electric fences to lose power, rendering the protective boundary between man and dinosaur useless. The humans become prey to these enormous predators as they roam freely hunting for human meat.

Two sequels were made, *The Lost World: Jurassic Park* (1997) and *Jurassic Park III* (2001).

Jeff Goldblum, Richard Attenborough, Laura Dern and Sam Neill inspect dinosaur eggs in *Jurassic Park* (1993).

OTHER WORKS INSPIRED BY GENETICS AND CLONING

Blade Runner

DIRECTOR RIDLEY SCOTT (1982)
STARRING HARRISON FORD, RUTGER HAUER,
SEAN YOUNG

Set in the dystopian nightmare of Los Angeles in 2019, Deckard (Harrison Ford) is a former blade runner. His job was to hunt down genetically-engineered humanoid robots known as replicants. Because of a bloody mutiny in colony where they are produced, their use is banned on Earth. Some replicants, however, try to defy the ban and return to Earth to have their life expectancy increased illegally. Deckard is recalled from retirement by the state to take on one final mission. He is ordered to terminate four elusive replicants who hijacked a ship in space and have returned to earth seeking their maker. As Deckard's pursuit of the genetically-engineered renegades grows more and more violent, the dividing line between man and machine becomes blurred.

Twins

DIRECTOR IVAN REITMAN (1988)
STARRING ARNOLD SCHWARZENEGGER,
DANNY DEVITO, KELLY PRESTON

Not all movies concerned with genetics and cloning are scary and based in the realm of science fiction. On his 35th birthday, Julius discovers he has a twin brother and resolves

Three clones of Michael Keaton in *Multiplicity* (1996).

to track him down. When he finds Vincent it is hard to believe they are twins. Physically and mentally they are complete opposites. Julius is physically perfect and highly intelligent, Vincent is somewhat smaller and a crook. However, after time together they begin to discover the truth behind their conception as part of a genetics experiment. One baby had inherited all the 'good' genes, and the other inherited the 'genetic trash'. Guess which one is which.

Multiplicity

DIRECTOR HAROLD RAMIS (1996)
STARRING MICHAEL KEATON,
ANDIE MACDOWELL, ZACK DUHAME

Construction worker Doug Kinney's chaotic life changes when he befriends scientist Dr Owen Leeds who can successfully create human clones. Dr Leeds offers to make clones of Kinney to ease the pressure at work and home. At first, the clone proves useful. But then another clone is developed from the original clone, and then a third from the second. Each clone loses a degree of intelligence when copied and cause Kinney's life to descend into more chaos than before.

Gattaca

DIRECTOR ANDREW NICCHOL (1997)
STARRING ETHAN HAWKE, UMA THURMAN,
JUDE LAW

Technology is used to genetically control the breeding of the population, with society divided between the genetically perfect 'valids' and the naturally-conceived 'in-valids'. Those in positions of power discriminate in favour of the valids. The in-valids have a short life span, are susceptible to illness and are of limited intelligence and are always given the menial jobs. An in-valid, Vincent (Ethan Hawke), wants to be an astronaut, but due to genetic discrimination has no chance. He realizes to get anywhere he has to cheat the system and use a valid's DNA to get through the interview. He buys the DNA from former athlete Jerome (Jude Law), who is now paralyzed from the waist down. Samples of Jerome's blood, hair and urine get Vincent accepted through the system of the Gattaca Aerospace Corporation. As the launch date for his mission to Saturn's moon Titan approaches, Vincent struggles to maintain his 'valid' identity as the net closes in on him.

The Stepford Wives

DIRECTOR FRANK OZ (2004)
STARRING NICOLE KIDMAN, BETTE MIDLER,
MATTHEW BRODERICK

A remake of the 1975 movie, Joanna is a career-driven executive producer on a reality television show, and Walter is her long-suffering husband. After a major disaster at work, Joanna suffers a nervous breakdown and Walter insists they move from Manhattan to Stepford to recuperate. Once there, they find that every aspect of living in suburbia is suspiciously perfect, and that all the averagely attractive men have incredibly good-looking, well-groomed and subservient wives. They discover that the women are clones of themselves, with a brain-implanted microchip to control their behaviour.

ROBOTS AND MACHINE MEN

I, Robot

NOVEL BY ISAAC ASIMOV (1920 – 1992)

An American author and professor of bio-chemistry, Isaac Asimov is best known for his works of science fiction and is widely considered a master of the genre. Along with Arthur C. Clarke and Robert Heinlein, Asimov is regarded as one of the three most influential science fiction writers of the 20th century.

One of Asimov's most famous works is the Robot series of novels and short stories. *I, Robot* is a collection of nine science fiction short stories first published in 1950. All Asimov's robots follow the Three Laws of Robotics which are programmed into each robot's brain. The Laws are designed to ensure that the robot does not turn against its creator. Asimov believed the Laws were his most enduring contribution to science fiction.

THE THREE LAWS OF ROBOTICS

1) A robot may not injure a human being or, through inaction, allow a human being to come to harm.
2) A robot must obey the orders given to it by human beings, except where such orders would conflict with the First Law.
3) A robot must protect its own existence as long as such protection does not conflict with the First or Second Laws.

Officer Del Spooner (Will Smith) looking for a killer among an army of robots in *I, Robot* (2004).

The Three Laws have had a huge impact on the evolution of robot stories over the years. Science fiction writers, like their fictional protagonists, prefer to keep control of their robots, giving them boundaries, limiting their actions and remaining at all times superior beings.

I, Robot

Director Alex Proyas (2004)
Starring Will Smith, Bridget Moynahan, Bruce Greenwood

Set in Chicago 2035, a world where robots are programmed to obey Asimov's Three Laws. The robots live and work peacefully alongside humans. Del Spooner, a technophobic cop investigates a case of alleged suicide, but his suspicions lead him to believe a robot has broken one of the Three Laws and may be involved in murder. If one robot can commit a crime how many more are capable of the same thing....

OTHER WORKS INSPIRED BY ROBOTS

Metropolis

Director Fritz Lang (1927)
Starring Brigitte Helm, Alfred Abel, Gustav Frölich

In the futuristic city of Metropolis, society is divided into two classes. The managers live in luxurious skyscrapers and the workers live and toil underground. The city was founded and built by the autocrat businessman Joh Fredersen. He meets up with his old collaborator, scientist Rotwang who now lives in the lower levels of the city. The two were once friends but argued over the love of a woman and Rotwang now bears an angry grudge. He also wears a mechanical hand and has invented a robot to do his evil bidding. Rotwang forces a beautiful girl, Maria, to give Machine-Man her face and transforms the robot into an evil double of Maria. He then commands the killer robot to destroy Fredersen and his city. Brigitte Helm is stunning as both Maria and her psycho-clone in this classic silent film.

The Day the Earth Stood Still

Director Robert Wise (1951)
Starring Michael Rennie, Patricia Neal, Hugh Marlowe

A flying saucer orbits Earth and lands in Washington DC in the Cold War era. A lone alien humanoid occupant disembarks to communicate peacefully with the citizens but is immediately attacked. Gort, a large powerful bodyguard robot programmed to respond to violent aggression steps in. Gort is a member of a race of super-robot enforcers invented to maintain intergalactic peace and programmed to destroy the Earth if provoked.

Silent Running

Director Douglas Trumbull (1972)
Starring Bruce Dern, Cliff Potts, Ron Rifkin

On earth, all the trees have long vanished. Freeman Lowell looks after plants in a giant space-craft greenhouse. When orders come from earth to destroy the greenhouse, Lowell can't go through with it. He murders his crewmates and changes course for the Sun accompanied only by three robots – Huey, Dewey and Louie. He teaches them to play poker, plant trees, and bury the dead.

Star Wars

DIRECTOR GEORGE LUCAS (1977)
STARRING MARK HAMILL, HARRISON FORD, CARRIE FISHER

A classic movie that shaped American cinema with action, romance, dwarves in bear suits, alien creatures and two of the most famous robots of all time: mechanic droid R2-D2 and protocol droid C-3PO. R2 is capable of many tasks, and while he cannot speak in a language the audience can decipher, he communicates in a series of beeps which can be understood by other droids. C-3PO is a loyal and highly intelligent robot, and due to his many decades as a protocol droid, has become anxious and prone to worrying: 'Did you hear that? They shut down the

main reactor. We'll be destroyed for sure. This is madness'.

Star Wars spawned dozens of books, two sequels, one prequel, two TV movies, comic books, action figures and millions of fans.

Douglas Adams' The Hitchhikers Guide to the Galaxy

BBC TV (1981)

Amongst all the unusual characters aboard the spaceship SS *Heart of Gold*, is a robot named Marvin, the Paranoid Android. Marvin claims to be 50,000 times more intelligent than a human, and due to his enormous intellect he suffers from severe depression. He often complains about being depressed and answers questions sarcastically, frustrated at having a 'brain the size of a planet' and not ever being sufficiently challenged to use it. His catchphrases include, 'Life, don't talk to me about life', and 'I think you ought to know I'm feeling very depressed'. Always

R2-D2 played by Kenny Baker and C-3PO played by Anthony Daniels in *Star Wars* (1977).

spoken in the same dreary, monotonous tone.

A British movie version of *The Hitchiker's Guide to the Galaxy* starring Martin Freeman was released in 2005, directed by Garth Jennings.

The Terminator

DIRECTOR JAMES CAMERON (1984)
STARRING ARNOLD SCHWARZENEGGER, LINDA HAMILTON, MICHAEL BIEHN, PAUL WINFIELD

In post-apocalyptic 2029, machines are systematically destroying what is left of mankind. The Terminator is a humanoid cyborg assassin who is sent back in time to 1984 to kill Sarah Connor. Her as yet unborn son, John, in the future will lead a resistance against the robot uprising. Hot on the Terminator's trail is a soldier, Kyle Reese, a man sent to protect Sarah, in order to ensure the future existence of her son and give humanity a chance at survival against the machines.

Sequels: *Terminator II Judgement Day* (1991), *Terminator III Rise of the Machines* (2003), *Terminator Salvation* (2009)

Short Circuit

DIRECTOR JOHN BADHAM (1986)
STARRING ALLY SHEEDY, STEVE GUTTENBERG, FISHER STEVENS

A former military robot, Johnny 5, is struck by lightning bolt and malfunctions. As a result he is technically altered becoming intelligent and curious. As witnessed in many previous works going all the way back to *Frankenstein*, the classic device of the lightning bolt strikes again as a re-animation trigger.

RoboCop

DIRECTOR PAUL VERHOEVEN (1987)
STARRING PETER WELLER, NANCY ALLEN, DAN O'HERLIHY

Detroit in the future – crime is out of control. The police are fighting a losing battle and Officer Murphy is brutally gunned down by a bunch of thugs. Using cyber-technology his body is reconstructed within a steel shell and named RoboCop. Seen as the future of law enforcement, the powerful cyborg is out to fight crime, but the fight takes him places he doesn't want to go including into the memories of his haunted past.

Sequels: *RoboCop II* (1990), *RoboCop III* (1993), *RoboCop:Prime Directives* (TV series 2001)

Universal Soldier

DIRECTOR ROLAND EMMERICH (1992)
STARRING JEAN-CLAUDE VAN DAMME, DOLPH LUNDGREN, ALLY WALKER

Two US soldiers kill each other in Vietnam. Listed as MIA, they are frozen and shipped to a US army secret facility for re-animating dead soldiers. A team of scientists turn the two into cyborg super-soldiers known as UniSols. They become part of an elite fighting force to combat terrorism.

A.I. Artificial Intelligence

DIRECTED BY STEVEN SPIELBERG (2001)
STARRING HALEY JOEL OSMENT, JUDE LAW, FRANCES O'CONNOR

Set in the year 2104, Henry Swinton, an employee of the company Cybertronics, is selected to take home a prototype child android known as a 'mecha'. David is built to look identical to a human boy. The Swinton's son Martin does not accept David's place in the family and sibling rivalry ensues.

Iron Man

Director Jon Favreau (2008)
Starring Robert Downey Jr,
Gwyneth Paltrow, Terrence Howard

Tony Stark's company sells weapons of war. He is captured and wounded in Afghanistan and his captors want him to make missiles for them. Instead Stark creates an armoured suit and escapes. Back in the US he announces his company will cease making weapons and

he makes an updated armoured suit in which he returns to Afghanistan.

Surrogates

Director Jonathan Mostow (2009)
Starring Bruce Willis, Radha Mitchell,
Ving Rhames

Set in a futuristic world where people live their lives via robot surrogates. It's a world where there is no crime, pain or fear. A murder is committed, the first one in years. FBI agent Greer uncovers an evil conspiracy behind the surrogates and risks his life to unravel the mystery.

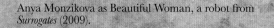

Anya Monzikova as Beautiful Woman, a robot from
Surrogates (2009).

EXTREME EXPERIMENTS

The Strange Case of Dr Jekyll and Mr Hyde

BY ROBERT LOUIS STEVENSON (1850–1894)
Published in 1886, this spine-chilling thriller recounts the story of Dr Henry Jekyll who believes that everyone has a good side and a bad side. To test his theory Jekyll concocts a drinking potion which transforms him into the evil Mr Edward Hyde who is a manifestation of the repulsive side of Jekyll's personality. Despite being the same person 'inside', Jekyll is transformed physically into Hyde who is a demented violent madman. Jekyll must face the reality that this horrific darkness within him is something he did not know existed.

Stevenson's classic story is one of the duality of man, the restrained monster within us that, if unleashed, can have devastating results. The term 'Jekyll and Hyde' is now used universally to describe split personality disorders, or people who are subject to violent mood swings behaving differently from one day to the next.

Dr Jekyll and Mr Hyde at the Movies

Filmed many times it continues to disturb audiences through the power of its narrative and its underlying psychological implications. The three earliest releases feature John Barrymore (1920), Fredric March (1931) and Spencer Tracy (1941) as the metamorphosing serial killer Jekyll/Hyde. Many critics believe the silent movie of 1920 to have the best special effects, whilst Spencer Tracy in the 1941 version is recognized as having the most menacing presence as Mr Hyde.

The plot in all the versions remains the same, following Stevenson's original storyline: Jekyll succeeds in his experiments to split his good and evil sides with chemicals and transforms into Hyde to commit horrendous crimes.

1920 VERSION

DIRECTOR JOHN S. ROBERTSON
STARRING JOHN BARRYMORE, CHARLES LANE, BRANDON HURST
This movie, a silent classic, has amazing special effects. It is hard to believe that the same actor plays Dr Jekyll and Mr Hyde. The make-up, the lighting, and excellent acting give a true impression of a different, darker, more evil man. It's still scary after all these years!

1931 VERSION

DIRECTOR ROUBEN MAMOULIAN
STARRING FREDRIC MARCH, MIRIAM HOPKINS, ROSE HOBART
The story retold starring Fredric March.

1941 VERSION

Director Victor Fleming
Starring Spencer Tracy, Ingrid Bergman, Lana Turner
The story once more starring Spencer Tracy.

SOME OTHER WORKS INSPIRED BY EXTREME SCIENCE

The Island of Dr Moreau

Novel by H.G. Wells (1896)

Edward Prendick is shipwrecked at sea and rescued by a passing boat. He arrives on a beautiful island, blissfully unaware that he has entered the sinister kingdom of Dr Moreau. Prendick is introduced to Dr Moreau and grows curious as to what his work on the island involves, especially as he is aware of the imported animals which travelled on the boat with him. He soon discovers that what goes on in this alleged paradise is extreme, bizarre and ghastly experiments on animals, and that Dr Moreau is an evil sadist, selfishly playing God in an attempt to create hybrid human-animal beings.

The Island of Lost Souls

Director Erie C Kenton (1932)
Starring Charles Laughton, Bela Lugosi, Richard Arlen

The Island of Dr Moreau was adapted into a film entitled *The Island of Lost Souls* and released by Paramount Pictures with an all star cast. The doctor is still a whip-cracking task master to a growing population of his own gruesome human-animal experiments, but he does have one prize result, Lota the beautiful panther woman.

The natives are restless in *The Island of Doctor Moreau* (1977).

Remakes of *The Island of Dr Moreau* were released in 1977 with Burt Lancaster and 1996 with Marlon Brando and Val Kilmer.

The Invisible Man

NOVEL BY H.G. WELLS (1897)

A victim of his own experiments, the mysterious Dr Griffin, new to the village of Iping, is always seen swaddled in bandages, wearing dark goggles, a hat and coat. He is reclusive and rude to those that try to talk to him. The locals notice that every inch of him is covered in material, and that he spends all day alone working with his scientific apparatus, only ever venturing out at night. Dr Griffin had discovered the secret of invisibility, and unable to restore his visibility he begins to descend into madness. In his pursuit of an antidote, Griffin goes on an invisible crime spree until eventually meeting his fate at the hands of the locals, his visibility returning as he draws his last breath.

The Invisible Man

FILM DIRECTED BY JAMES WHALE (1933)

STARRING CLAUDE RAINS, GLORIA STUART, WILLIAM HARRIGAN

The story was retold in a film considered one of Universal's great horror films of the 1930s. It spawned a number of sequels, plus many spin-offs using the idea of an invisible man that were largely unrelated to Wells' original story commencing with *The Invisible Man Returns* (1940). Several TV adaptations have been run including one starring David McCallum in 1975.

The Nutty Professor

DIRECTOR JERRY LEWIS (1963)

STARRING JERRY LEWIS, STELLA STEVENS, DEL MOORE

Science fiction and comedy meet when to improve his social life, a nerdish professor drinks a potion that temporarily turns him into the handsome, but obnoxious, Buddy Love. More recently, the film was remade in 1996 and starred Eddie Murphy in multiple roles.

A Clockwork Orange

DIRECTOR STANLEY KUBRICK (1971)

FROM A NOVEL BY ANTHONY BURGESS

STARRING MALCOLM MCDOWELL, PATRICK MAGEE, MICHAEL BATES

Set in an ultra-violent England of the future, vicious hooligan Alex DeLarge is convicted of murder and rape. While in prison, Alex learns of a government experiment in which convicts are brainwashed to reject violence. If he goes through with the program his sentence will be reduced. Sure that he can beat the aversion therapy and get out of jail quickly, Alex volunteers. But nothing can be that simple... .

One Flew Over The Cuckoo's Nest

DIRECTOR MILŎS FORMAN (1975)

STARRING JACK NICHOLSON, LOUISE FLETCHER, WILLIAM REDFIELD

ADAPTED FROM THE NOVEL BY KEN KESEY (1962)

Chief Bromden, half Native American, whom the authorities believe is deaf and dumb tells the story of a mental institu-

The Invisible Man was one of the team of extraordinary figures with legendary powers that adventurer Allan Quatermain assembled to battle the technological terror of a madman known as 'The Fantom' in *The League of Extraordinary Gentlemen* (2003).

tion ruled by the iron lady, Nurse Mildred Ratched. Into this grey world comes firecracker McMurphy, a brawling gambling man who wages total war with Nurse Ratched on behalf of his fellow 'patients'. McMurphy and the Chief become friends and end up fighting with the orderlies. They are taken to the 'shock shop' and given electro-convulsive therapy.

McMurphy staggers back to the ward zombie fashion but quickly explodes into laughter having fooled the other inmates. Nurse Ratched grows increasingly impatient with McMurphy's trouble making and sends him back to the shock shop for further more-intensive electrotherapy...

The Incredible Hulk

(TV Series 1978–1982)
Starring Bill Bixby, Lou Ferrigno, Jack Colvin

David Banner is a brilliant scientist but when a lab experiment goes badly wrong his body undergoes a transformation into 'The Incredible Hulk'. The Hulk is about seven feet tall, hugely muscular and powerful, and bright green. After usually destroying whatever threatens Dr Banner, the Hulk changes back to normal. Lou Ferrigno was the angry, but sensitive, Hulk. This much-loved TV show re-invigorated superheroes who were in the doldrums.

Movie remakes: *Hulk* (2003), *The Incredible Hulk* (2008)

Bill Bixby morphing into Lou Ferrigno in
The Incredible Hulk TV series.

Jeff Goldblum in *The Fly* (1986). It is widely believed that the movie is loosely based on Franz Kafka's 1912 story *The Metamorphosis*, in which a man wakes from a nightmare to find himself transformed into a giant insect.

The Fly

DIRECTOR DAVID CRONENBERG (1986)
STARRING JEFF GOLDBLUM, GEENA DAVIS,
JOHN GETZ

Seth Brundle, a genius scientist is working on teleportation but begins to transform into a giant man/fly hybrid after one of his experiments goes horribly wrong. Things get even more complicated when, as Brundle continues to deteriorate, his girlfriend reveals she is pregnant. Fearing mutant offspring, she wants to abort the baby, but Brundle, now fully transformed into the absolutely grotesque 'Brundlefly', has other ideas.

The Island

DIRECTOR MICHAEL BAY (2005)
STARRING SCARLETT JOHANSSON,
EWAN MCGREGOR, DJIMON HOUNSOU

Lincoln and Jordan are best friends living in a repressive regime on a contaminated planet, where everyone is waiting for a lottery win. The prize is to move to a paradise island, the last uncontaminated spot in the world. But they find out that everything is a lie. All of the inhabitants are human clones kept as a source of replacement parts. Lincoln and Jordan make a daring escape pursued by sinister forces of the regime.

The Human Centipede

DIRECTOR TOM SIX (2009)
STARRING DIETER LASER,
ASHLEY C. WILLIAMS, ASHLYNN YENNIE

Two American girls are kidnapped and imprisoned in a terrifying basement along with a Japanese man. A retired German surgeon mutilates the trio of tourists and reassembles them into his sick lifetime fantasy, the human centipede.

Jerry Lewis and Stella Stevens in *The Nutty Professor* (1963). Legend has it that the Nutty Professor's sleazy alter ego, Buddy Love, was a satirical swipe at Jerry Lewis' long-time Hollywood screen partner, Dean Martin.

Postscript:

ALZHEIMER'S DISEASE AND ELECTRIC SHOCK THERAPY

Deep brain stimulation (DBS) has now been used successfully in tens of thousands of patients with Parkinson's as well as having an emerging role in Tourette's Syndrome and depression. The mild electric stimulation used in DBS is not to be confused with electroconvulsive therapy (ECT), or 'shock therapy', which has itself been proved to be a valuable technique for helping some patients with severe depression. DBS is now being tested on patients with Alzheimer's Disease.

In DBS, surgically implanted electrodes deliver a series of mild electric pulses to the affected region of the brain. Brain shrinkage, declining function and memory loss had been thought to be irreversible. In Alzheimer's, the hippocampus is one of the first regions to shrink. Its function is to convert short-term memory to long-term memory. Damage leads to some of the early symptoms of Alzheimer's - memory loss and disorientation. By late stage Alzheimer's, brain cells are dead or dying across the whole of the brain.

DBS has been tested on six patients with Alzheimer's Disease at the University of Toronto. In two patients, the brain's memory hub, the hippocampus, reversed its expected decline and actually grew. Lead researcher Professor Andres Lozano reported their success. 'In patients with Alzheimer's, you would expect the hippocampus to shrink by 5 per cent on average in a year. After 12 months of DBS, one patient had a 5 per cent increase and another had an 8 per cent increase. However, these are very early days and a very small number of patients are involved.' Precisely how DBS works is still unknown.

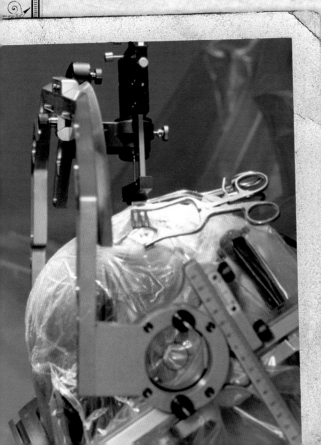

Insertion of a DBS electrode during surgery.

INDEX

This edition published in 2012 by
CHARTWELL BOOKS, INC.
A division of BOOK SALES, INC.
276 Fifth Avenue Suite 206
New York, New York 10001
USA

Reprinted 2014

© 2012 Oxford Publishing Ventures Ltd
Canary Press
An imprint of Oxford Publishing Ventures Ltd
Spring Hill House, Spring Hill Road
Begbroke, Oxford OX5 1RX
England

Although every effort has been made to trace and contact people mentioned in
the text for their approval in time for publication, this has not been possible in
all cases. If notified, we will be pleased to rectify any alleged errors or omissions
when we reprint the title.

ISBN-13: 978-0-7858-2879-2
ISBN-10: 0-7858-2879-6

Printed in China

PICTURES CREDITS: Internal Images: The Kobal Collection: 7, 158 © Universal / The Kobal Collection |
160 © Clover Prods / The Kobal Collection | 164 © Filmax / The Kobal Collection | 166 © MGM / The Kobal
Collection | 169 © Amblin/Universal / The Kobal Collection | 170 © Thinkfilm / The Kobal Collection | 173 ©
Columbia / The Kobal Collection | 175 © 20th Century Fox / The Kobal Collection / Digital Domain | 177 ©
Lucasfilm/20th Century Fox / The Kobal Collection | 179 © Touchstone Pictures / The Kobal Collection | 181
© Aip / The Kobal Collection | 183, 185 © 20th Century Fox / The Kobal Collection | 184 © Universal TV /
The Kobal Collection | 186 © Paramount / The Kobal Collection. **Alamy Limited:** 16, 30, 34, 65 © Mary Evans
Picture Library / Alamy | 40 © Roland Bouvier / Alamy | 50 © RIA Novosti / Alamy | 83 © The Print Collector /
Alamy | 128 © The Protected Art Archive / Alamy | 132 © STOCKFOLIO® / Alamy | 167 © Jeremy Pembrey /
Alamy | 172 © Pictorial Press Ltd / Alamy. **Corbis Images:** 21, 29, 86 © Bettmann/Corbis | 52 © Hulton-Deutsch
Collection/Corbis | 72 © Barnabas Bosshart/Corbis | 76 © Roger Ressmeyer/Corbis | 142 © Remi Benali/Corbis.
Bridgeman Art Library: 26 © Courtesy of Swann Auction Galleries. **TopFoto:** 39 © Print Collector / HIP /
TopFoto | 45 © RIA Novosti / TopFoto | 69 © 2005 TopFoto | 96 © The Granger Collection / TopFoto | 121 ©
ullsteinbild / TopFoto | 147 © Lightroom / NASA / TopFoto. **Getty Images:** 56 © Dan McCoy - Rainbow | 62, 92,
104, 109, 140 © Time & Life Pictures/Getty Images | 85 © Image Source | 90 © Photoservice Electa | 94 © Dean
Conger | 98, 125, 149 © Getty Images | 141 © Fox Photos | 146 ©NASA. **Mary Evans Picture Library:** 70 ©
Mary Evans / Sueddeutsche Zeitung Photo | 81 © Mary Evans Picture Library.